# Wheal Trewavas

Peter Joseph & Alasdair Neill

TAMAR MINES 2007

THE TREVITHICK SOCIETY KOWETHAS TREVITHICK

**Published by the Trevithick Society**
**for the study of Cornish industrial archaeology and history**



**ISBN 978 0 993502 15 6**

**The Trevithick Society is a Registered Charitable Incorporated Organisation, No. 1159639**



**Printed by R. Booth Ltd**
**The Praze, Penryn TR10 8AA**

**Typeset by Peninsula Projects**
**c/o PO Box 62, Camborne, Cornwall TR14 7ZN**

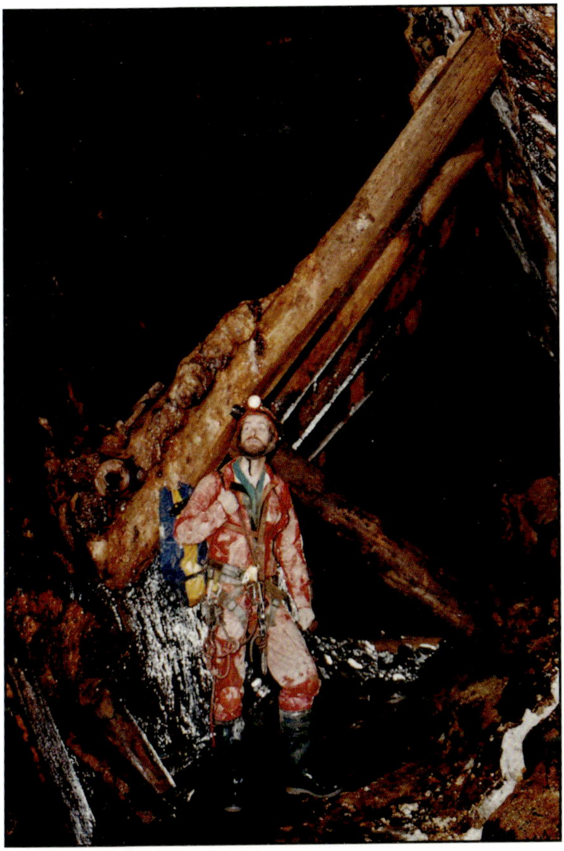

**Ali Neill in a stope in Devon Great Consols.**
**Photo: Paul Deakin**

The wild character of the coast is inferior to nothing of the kind in Cornwall; and *Wheal Trewavas*, which has engines perched on the cliffs, and its workings under the sea, is quite as picturesque as *Botallack* or *Levant* in the *St Just district*.

W. J. Henwood, 1843. On the Metalliferous Deposits of Cornwall and Devon, *Transactions of the Royal Geological Society of Cornwall* volume 5 p44.

# NAMHO 2007

This publication is in part funded by the proceeds of the 2007 National Association of Mining History Organisations annual conference which was held at Morwellham Quay in the Tamar Valley. As conference chairman it falls to me to write a few explanatory words by way of introduction

Whilst the conference was hosted by the Tamar Mining Group a number of other groups and individuals were directly involved in making the conference a success. Not least of these groups was the Plymouth Caving Group whose members, including the co-author of this book Alasdair Neill, ran many of the underground field trips arranged for the delegates. In addition to field trips we also ran a parallel lecture programme, the intention being to publish transactions at a later date. For various reasons the transactions never happened and the money which had been ring fenced for their publication remained unspent. When it became apparent that the transactions were not going to appear the decision was taken that the money would either be spent on an appropriate project in the Tamar Valley or, failing that, donated to an organisation with broadly similar aims.

In 2017 Robert Waterhouse finally finished his book on the Tavistock Canal and it was felt that it was a worthwhile project to support financially. The proposal was considered and agreed by various individuals and representatives of groups who had been involved in the 2007 conference. Discussing the idea with Dave Warne, the Chairman of the Plymouth Caving Group, he agreed that we should make a financial contribution to the Canal book and also suggested that an equal share should be donated to support this publication as a commemoration of the life and work of our friend Ali Neill who died tragically young in 2016. This seemed a wholly appropriate use of the money given Ali's lifetime contribution to mine exploration and mining history both in the Tamar Valley and further afield. It is doubly appropriate that this book receives this support given Ali's active involvement in the 2007 NAMHO conference.

On a personal level it is a privilege to be involved, albeit in a very small way, in bringing Ali's work to a wider audience, although a privilege tinged with more than a degree of sadness. I counted Ali as a friend for nearly twenty years, sharing mutual interests in both mine exploration and mining history. Over the years I spent many happy hours underground with Ali, often cursing his notoriously tight rigging (that's one thing that I won't miss). As a mining historian I regularly had recourse to Ali's encyclopaedic knowledge based on his extensive and solid research. Certainly my volume on Devon Great Consols would have been much the poorer without Ali's numerous contributions. Ali's great strength was a willingness to share his research unselfishly with others. If one picks up any book dealing with mining in South West England published in the last couple of decades the chances are that Ali's name will appear somewhere in the acknowledgements; this is his lasting legacy.

He is sorely missed.

Rick Stewart FSA
Calstock, November 2017

# Alasdair Neill, an Appreciation

Alasdair was born to John and Ruth Neill in Shrewsbury on 9th April 1960. John and Ruth were accomplished climbers and his mother said that Ali was climbing before he was born. Ali's pedigree is well founded.

Ruth Neill researched and wrote one of the first Cornish mining articles to be published: *The Carn Galver Mine* at Bosigran* (the old Count House being the Climbers Hut) published in *The Climbers' Club Journal* in 1962 and in the following year Ruth compiled and published *The Copper Mines of Snowdon*, again in the *Climbers' Club Journal*.

Ali naturally became accomplished in many outward bound activities: canoeing, sailing, climbing and caving. He existed on minimal personal comforts, preferring to spend his time either researching caves or mines anywhere they may exist on the planet I know; I am the custodian of his work, or exploring and surveying these dark places.

His supporting role of providing data and survey maps enabled the compiling of many articles and books including statistical tomes of work such as those of Professor Roger Burt. His surveys of the ever growing extensive cave complex of Matienzo, Spain will be another worthy legacy.

His funeral, which was held at Plymouth Crematorium, was respectfully unusual in that six friends carried him on a rescue stretcher; we were dressed in caving attire and the coffin was adorned with his helmet, lamp, measuring tape and family flowers. The service included joyful antics and reminiscences of his life.

Finally a good gathering of cavers and family did a 'farewell to Ali' trip at Wheal Coates on 23rd December 2016 when Storm 'Barbara' (St Barbara is the patron saint of miners) graced us with her presence.

Kevin Baker a climbing/caving friend of 30 years

***Ali also wrote an article on Carn Galver Mine, published in the *Journal of the Trevithick Society* No. 29, 2002.**

# Author's Foreword

In 2012 I was talking to Alasdair (Ali) Neill and mentioned that I had been researching Wheal Trewavas and was contemplating writing a book on the mine. So was he, it seemed, so I let him lead on, but continued with my own work in the background. We didn't discuss the mine again until 2016, when I heard that he was - finally - working on a manuscript, so I emailed him in August that year. Ali replied saying that he had been doing a lot of work on the mine over the previous month to get his manuscript completed as far as he could as he had been diagnosed with advanced lung cancer. This was not good news. A month later he sent me a copy of his manuscript for my interest. I made a few comments and sent him a copy of my notes. The last email I had from him was on August 26th when he sent some underground images for me to try to tidy up. I knew that Ali was being treated for his cancer; he had been invited to my surprise 60th birthday get-together at Levant on September 16th but had been too unwell to attend. Unfortunately, Ali died in Plymouth on October 28th.

Ali was an exceptional researcher; his ability to ferret out information which had few, if any, equals. He was always happy to share research and point others in the right direction; many researchers have benefited from his experience. Ironically, when he and I were talking in the early part of 2016, I mentioned that as he was the person who knew where to look for information it would be a good idea if he wrote something on the topic. He replied that others had already suggested this! Ali was not a person for whom publishing was important; his thrill was in the chase. Hopefully, though, he won't mind his name being associated with this publication.

I am grateful to Adam Sharpe for various helpful discussions on the layout of the mine and to him and Graham Thorne for proof-reading the manuscript.

Pete Joseph

# Introduction

Perched on the cliffs 3km to the west of Helston, the two engine houses of Wheal Trewavas appear to be part of the landscape, almost as if they have grown out of the granite on and out of which they were built. This mine is very well known, in that there are many references to it, notably on the World Wide Web, however the actual amount of historical information available is very small; sadly this is typical of so many Cornish mines. The one common misconception is that the mine closed after the sea broke in – there is absolutely no evidence to support this claim whatsoever, only that the mine flooded after it closed.

Certainly the two engine houses here are as dramatically situated as those of the Crowns section of Botallack Mine and at Levant Mine, although at the latter the two built lowest down and nearest the sea were demolished by about 1860. While there are no mines on the coast of Mounts Bay which were in any way comparable with those of St Just, in terms either of production or of longevity, Wheal Trewavas is a testament to the ingenuity of the Cornish miner and engine house constructor. Nowhere else in Cornwall were such large engine houses constructed adjacent to such precipitous drops and nowhere else around Mounts Bay did mines extend out to sea. This book attempts to demystify this history and collate all the known information in a single narrative for the first time.

# Chapter 1

## The History

A Wheal Trewavas seems to have been working by the early 1820s, and in 1825 the Cornwall South Consolidated Copper and Tin Mining Company was formed in order to acquire 'the extensive Mines of' Wheal Hendra, Wheal Trewovas (Trewavas) and Wheal Tregennebus (Tregonibris).[1] It should be noted though that there is a possibility of confusion because of the presence of a Trewavas tin stream south of Helston.

> The Mines abound with Tin Lodes and Streams of the richest description, and from the appearance of their surface, particularly on Wheal Hendra and Trewovas, it bears undoubted evidence of its being one of the principal districts from whence the ancient Phoenicians procured their immense quantities of fine Tin, by Shallow Mining and Streaming alone, no other methods at that time being known.

Fourteen whim shafts and two engine shafts had been sunk and timbered and there was also an unused steam engine in its house with smiths' and carpenters' shops and new whims. Unfortunately the report gives no clear idea as to where any of these were located and they may have been on Wheal Hendra or, more likely, on more than one mine. About a third of the shares had been taken up by the adventurers in several of the adjacent mines, viz. Wheal Anne, Wheal Trevenen and Wheal Trumpet (see figure 1). The company had a capital of £62,500 in 2,500 £25 shares. The principal agent and mine surveyor was Thomas Teague of Redruth; the resident captain was John Martin. One of the auditors was John Sylvester of Helston, who will figure briefly but significantly in this narrative later. While no details are known of these early operations, the profit of £1,920 referred to by Watson in 1843 as having been made a 'in a former working' may refer to this, or even earlier operations.[2] The underground workings provide some evidence that mining was certainly not new to this site in the 1830s and that a considerable amount of work took place prior to 1834.

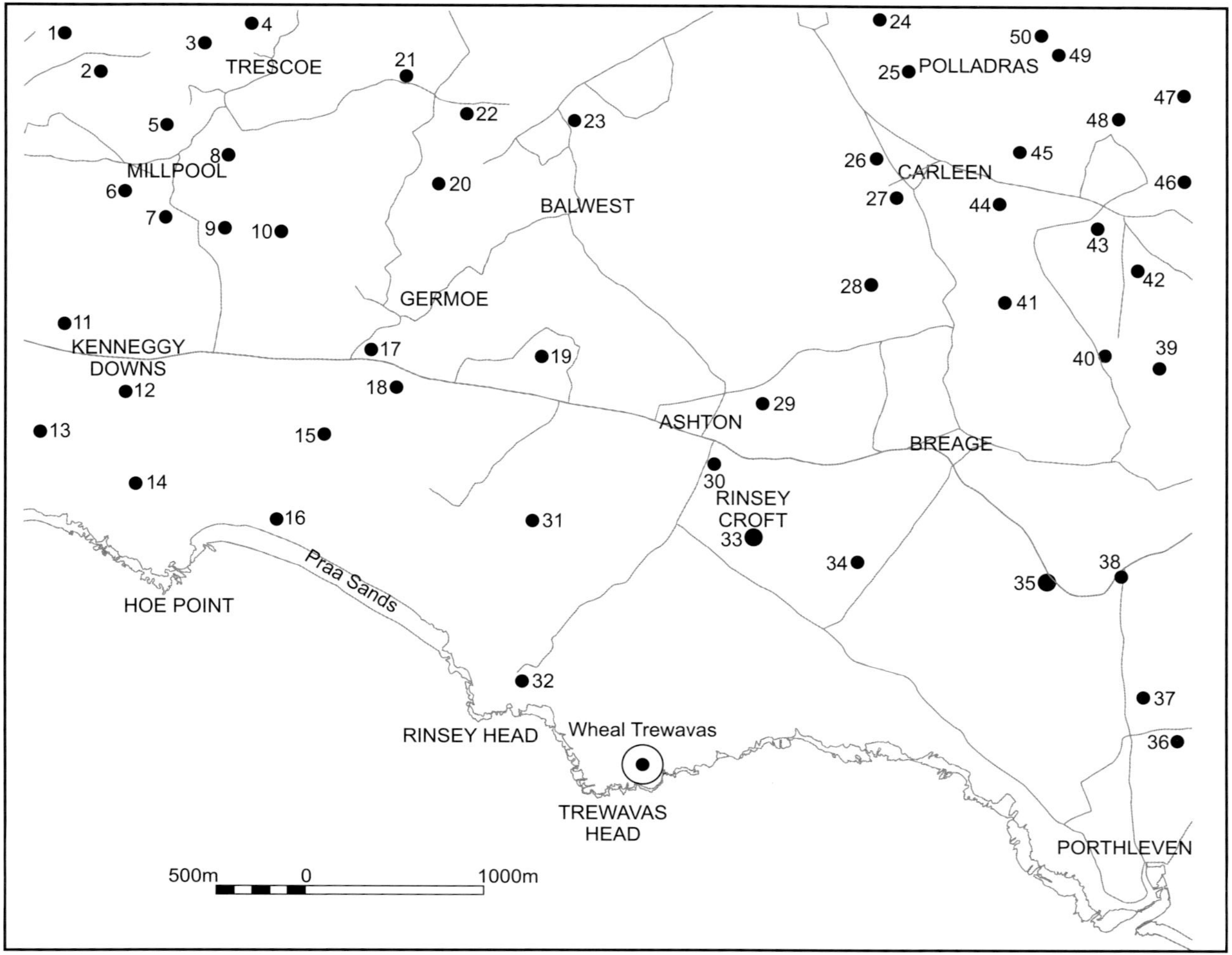
TRESCOE
MILLPOOL
KENNEGGY DOWNS
GERMOE
BALWEST
POLLADRAS
CARLEEN
ASHTON
BREAGE
RINSEY CROFT
HOE POINT
Praa Sands
RINSEY HEAD
Wheal Trewavas
TREWAVAS HEAD
PORTHLEVEN
500m
0
1000m
1
2
3
4
5
6
7
8
9
10
11
12
13
14
15
16
17
18
19
20
21
22
23
24
25
26
27
28
29
30
31
32
33
34
35
36
37
38
39
40
41
42
43
44
45
46
47
48
49
50

**Figure 1. Location of Wheal Trewavas and adjacent mines.**

**1. Retallack Mine**
**2. Croft Gothal Mine**
**3. Wheal Fortune**
**4. West Great Work Mine**
**5. Wheal Leeds**
**6. Millpool Mine**
**7. Black Down**
**8. Esconick**
**9. Bogill**
**10. Wheal Lemon**
**11. Great Western Mine**
**12. St Aubyn & Grylls Mine**
**13. Wheal Speed**
**14. Keneggy**
**15. Sydney Godolphin Mine**
**16. Sydney Cove Mine**
**17. Leeds and St Aubyn's Mine**
**18. Leeds and St Aubyn Mine**
**19. Wheal Grey**
**20. Lady Gwendoline**
**21. Trescove Tin Mine**
**22. Wheal Reeth**
**23. Great Work Mine**
**24. Polrose Mine**
**25. Polladras Mine**
**26. Wheal Leeds**
**27. Carleen Mine**
**28. West Wheal Metal**
**29. Trevena Mine**
**30. Wheal Ann**
**31. Hendra Mine**
**32. Wheal Prosper**
**33. Wheal Ellen**
**34. Wheal Mary**
**35. Wheal Rib**
**36. Wheal Saturn**
**37. Treeza Mine**
**38. Hansard Mines**
**39. South Providence**
**40. Wheal Fortune**
**41. Wheal Emma**
**42. Carnmeal Mine**
**43. Wheal Metal**
**44. Wheal Corn**
**45. Wheal Vor**
**46. Sithney Wheal Metal**
**47. North Metal Mine**
**48. Wheal Vreah**
**49. Gwin & Singer**
**50. Penhale Wheal Vor**

Wheal Trewavas was opened, or re-opened, late in 1834 by the Trewavas Mining Company, the name taken from the estate on which it was located, and some of the remaining shares in the mine were advertised in May and October 1835. The apparent speed with which the mine was developed suggests that it was restarted. This cost book company was initially divided into 64 shares, the amount of each not being known.

Cost book companies were common in Cornwall until the latter part of the 19th century (although Levant continued as a cost book company into the 20th century). These were essentially unlimited liability companies: when times were good the profits were distributed pro rata between the shareholders; unfortunately, when a mine made a loss, the loss was also distributed pro rata! These tended to be hand-to-mouth organisations which kept little, if any, funding aside for new equipment and exploration. Limited liability companies appeared in Cornwall quite soon after the introduction of the Company's Act, along with its associated limited liability for shareholders. Scrip companies were almost the complete opposite of cost book companies in that a scripholder had no liability beyond his or her financial obligation to the shares they had bought. On the debit side the scripholder had absolutely no say in how the company was run.

The mine was well-situated, its cliff-side location meaning that workings could be drained by the simple expedient of driving an adit into the cliff. The area was also well-provided with lodes, those within the sett including Way Sowan,* Nimble Cutter, North and South, all of which carried copper. These lodes all trend around south-east to north west; Great Tin Lode trends east to west in the northern part of the mine. In addition, several crosscourses existed, one of which (the Way Sowan Crosscourse) may have been mineralised. This crosscourse faulted the Way Sowan Lode, moving the part west of the crosscourse nearly 10m north (see figure 11).

The first detailed account of the mine seems to be a letter dated 3rd July 1835, from Thomas Teague to Noah Coward at Gunnislake Mine.[3] Teague was the purser and a principal adventurer in Tresavean, and Coward was involved with him there and in many other mines. Coward and Teague were adventurers in Trewavas. Teague wrote,

> I was yesterday underground at Trewavas Mine in an excellent course of ore in the 22 fm level East & a very kindly lode rising in the back towards a 15 fm level, so soon as this is holed, a considerable quantity of rich ores can be stoped down, the western levels are in hard ground & poor. They have a kindly lode in the shaft, the East End is moderate ground but I fear the Rock, if the ground keeps moderate by the lode I have no doubt of its doing well & there is a fine channel of ground east, they say dipping west which is more favourable. The ores for sale next week produces 18¾ but what they are now raising is much better.

A more general description appeared in the *Mining Journal* in October.[4] The mine was in the parish of Breage, on land owned by the Reverend Canon

***In a letter on lode intersections in the *Cornish Telegraph* of 25th November 1863 p3 John Truran refers to the roof of a 'small sowan, or zone (possibly an error for 'zawn') excavated by the sea'.**

Rogers. By this time the deepest level worked was the 22, where a level had been driven 10 fathoms west of the shaft and 40 fathoms east, beneath the sea.

The sea bed was apparently discovered by sounding the back of the level and was 14 fathoms above the level. In the 19th century 'sounding' would have meant the depth of the sea being measured by lengths of weighted rope thrown over the side of a boat but this does not appear to be the case here; one almost wonders if the miners were knocking the back of the level to try to determine its thickness. The lode in the eastern end was '3 to 4 feet big' about 20 fathoms under the sea; this appears to be in the lowest level. The agents were rather concerned about taking away too much of the 'backs' (the 'roof' of the level) for fear of 'an intruding visit from old Neptune'. The mine had been

> discovered by some of these amphibious creatures who obtain their livelihood by fishing in the summer and mining in the winter, who observed from their boats the lodes or mineral veins in the cliffs of this estate.

> Since that time

> a steam engine has been erected, and as the shaft has been sunk many fathoms on the lode in the nook in the cliff, it became necessary to excavate to a large extent to make room for the engine-house, horse-whim, &c., and which are now at full work, about 100 feet from surface. The corner of the boiler-house is only one foot from the edge of a further precipice of 150 feet, literally perpendicular.

From July 1835[5] increasing quantities of copper ore were being sold, usually bi-monthly, at the Cornish copper ore ticketings. For example, 112 tons were sold on 10th September at Redruth, and this realised £941 4s 6d.[6] Considering how early this was in the mine's history, and in view of the apparently limited extent of the workings, this was quite a tonnage. This was matched in November and exceeded, slightly, in December.[7]

D. B. Barton states that the pumping engine came from Harvey & Co. of Hayle, at a cost of £807, but that it was sold in September 1836.[8] This does not tally with the engine being on the mine in 1835 but could possibly relate to the final payment for the engine. It is not clear how the engine arrived at Trewavas however, given the generally poor state of all British roads at this time, it is most likely that the main part of the journey was by sea. Possibly the parts were delivered to the quayside in Porthleven and then taken by road, however it is also possible that the parts were landed at Praa Sands, and from there by road or track.

Even though the engine was a comparatively small one, the beam weighed 12½ tons and would have required a wagon and a team of at least a dozen sturdy horses to pull it. Depending on the sea conditions it would have been a comparatively simple task to haul the cylinder, bob and boilers up the cliff, which is about vertical here, although this would be an unusual system of delivery. In addition, each of the three boilers would have weighed on the order of 12 tons, measuring around 25 feet in length and six feet in diameter.

Building the engine house adjacent to the shaft and close to the cliff edge required the skills of stevedores and the ingenuity of structural engineers; frequently these were combined in one and the same person, the long experience in building houses and erecting engines allowing a certain amount of alacrity in the work. It is interesting to note too that modern health and safety regulations do not allow mine buildings to be constructed adjacent to mine shafts. In Cornwall many pumping engine houses, weighing on average 900-1,000 tons, have stood next to shafts without any harm for 150 years or more.

**Figure 2. Photograph showing what was required to transport a Lancashire boiler (location and date unknown) in the nineteenth century.**

The deep adit was 18 fathoms from surface, with three shallower adits above. This adit, the highest level to which water was usually pumped, was only about three feet above sea level, which must have caused many problems whenever a strong wind from the south or south-west was combined with a high tide. Possibly a timber cofferdam was used to protect the adit mouth when the weather was poor.

By the time of the report the engine shaft was down 32 fathoms. Because the road to the mine was so steep there was no possibility of taking coal down, or bringing copper up, by cart. Consequently a tram road had been cut into the rocks and was operated by a horse-whim; this implies that this applied to Old Engine Shaft. Another complication was feed water for the boiler, which had to be brought nearly half a mile in pipes along the cliffs from a spring. This probably refers to the spring to the west of the mine at SW 5957 2643. Because of the slope of the land to the south, the pipe ran along the cliffs for some distance before reaching the mine (there is some evidence for this below the dressing floors), a considerable feat of surveying and engineering. The cost of operations to date was given as £3,130 8s, the mine allegedly being worth six times that.

Shares in the mine continued to be advertised through late 1835 and early 1836, while in February 1836 it was reported that an important discovery was made of a new copper lode, cut six feet south from the 22 fathom level, in whole (unbroken) ground. Two feet wide, it was said to be worth £40 per fathom, an extravagantly high value for a copper lode.[9] The prospects of the other parts of the mine were said to be equal, if not beyond, 'the high expectations which have been formed'. By April the shares seem to have been subdivided as William Trenery, the Redruth share dealer, was offering 3-128th shares for sale.[10]

Another discovery was recorded in May, 'about the best ever made in the mine'.[11] It was reported that this lode could be traced nearly a mile towards Porthleven, running through the East Trewavas sett. This may refer to the Great Tin Lode, in the northern part of the mine (see figure 11), the only east-west lode thought to run through the sett. The same number of shares was advertised for sale in September, along with 4-200ths in East Trewavas.[12]

At the account meeting held in September 1836 the mine's profit of £640 was divided amongst the shareholders.[13] The remaining £970 was left in hand to contribute towards the cost of 'the erection of a powerful engine which is nearly completed'; this partly agrees with Barton, implying that the engine parts were delivered on the mine from 1835 and put to work the following year. During this year the mine employed a total of 161 persons, comprising 102 men, 42 women and 17 children.[14] A 1-128th share was advertised for sale in September 1836[15] while in October 2-265th shares sold for £300.[16] In November the North Lode was cut, and, from appearances, large quantities of copper ore would be returned from depth. According to the *Royal Cornwall Gazette*,

> This neighbourhood for a number of years past has been completely lost sight of, but from the evidence of the above mine, and many others in its immediate neighbourhood, such as Wh. Unity, Wh. Penrose, Wh. Vor, Wh. Fortune, and many others, the old adage may fairly be applied in this case, viz:- that there is little done and much to do, particularly in the parish of Breage.[17]

At the same time a large copper lode was discovered in East Trewavas, possibly the same lode mentioned above; here an adit was being driven inland from the cliffs in the hope of cutting this and many other alleged lodes at 30-40 fathoms below the surface.[18] During 1836 Wheal Trewavas sold 1,236 tons of copper ore for £11,625 8s 6d. Sales averaged over 200 tons per ticketing, sold bi-monthly. The following year production was similar at 1,156 tons for the somewhat lower price of £7,174 17s 6d, the difference being predominantly because of the lower price achieved per ton of copper ore.

In Cornwall and Devon the tin mines were ruled by Stannary Law, which defined what a miner could or could not do. The area from Dartmoor to Lands End was divided into stannaries, each of which had a stannary town. The Stannaries were ruled, if that is an appropriate term, by the Warden of the Stannaries Court, though his assistant, the Vice Warden, appears to have done all of the work. The Court worked as a law court and eventually handled all of the cases relating to all of the mines of Cornwall and Devon. To go with the law there were two stannary prisons, the one at Lydford in Devon having such a poor reputation it was said they would hang a man first then find him guilty afterwards. The towns were within tin producing areas and were where tin ore was sent to be smelted and sold. In 1837 the stannary laws were expanded to include copper mines; however the laws relating to mining were changed and there was little, if any, difference as to how the copper mines operated.

The account meeting on 21st November (for September and October) 1836 showed a profit of £640, leaving £1,500 with the purser; the prospects were 'truly encouraging'.[19] A dividend of £5 per share was declared, however 'a much larger sum would have been divided had there not been a very large outlay for engine, &c., amounting to upwards of £2000 within the last two months, which has now been cleared off'. A few more details at this period are reported from the proceedings of the West Cornwall Mines Investment Company,* a concern formed to purchase shares in mines, which had acquired a 1/64th share in Trewavas. A report of this company dated May 1837 stated that the 20 fathom level east end was worth £30-40 per fm, but despite these riches was suspended for being too close to the sea bed. The 32 east was worth £11 and a rise in the back worth £20. The 45, evidently the deepest level, showed a large and kindly lode, but only worth £5 per fathom. The authors of the report, John Sampson and Sampson Trevethan of Chacewater, concluded 'This mine has not the flattering appearance she had six months since'.[20]

The 1838 AGM of the West Cornwall Mines Investment Company showed that on the company's 1/64 share in Trewavas £30 in dividends had been received and the share was valued at £620.[21] This indicates a total dividend for the year of £1,920.

The pursers at this time were Collan Harvey and his son Richard, of St. Day, two members of one of the great dynasties of Gwennap copper mining.[22] However, that the financial management of the mine was not all it should have been is revealed in letters from Alfred Jenkin of Redruth to his employers, the copper smelters, Pascoe Grenfell and Sons. These show that on several occasions the mine delayed for a month or more on calling for payment for copper ores purchased at the ticketings by Pascoe Grenfell & Sons.[23] This was very unusual, the process of invoicing by the mines and payment by the copper smelters otherwise invariably following a very efficient timetable.

Copper ore sales continued at a low level through 1838 (989 tons) and 1839 (976 tons). It is likely the mine was making losses in these years, one source stating the loss in 1839 was £2,000.[24]

***For details of the West Cornwall Mines Investment Company see: Brooke, J., 1990. The First Investment Trust. *Journal of the Trevithick Society* No. 17 p.48-52.).**

**Trewavas Mine.**

TO BE SOLD, a Hundred and Twenty-Eighth Part in that valuable Mine.
Lowest price £250.
Apply to Mr. Penaluna, Helston, or Mr. Nettleton, Plymouth.
Dated 31st March, 1840.

**Figure 3. Advertisement for some very expensive shares in Wheal Trewavas, *Royal Cornwall Gazette* 3rd April 1840. Shares were rarely sold for this sort of price, except for mines such as Botallack, however the likelihood of actually getting this money back from dividends was optimistic at best.**

In 1843 W. J. Henwood published his 386-page work *On the Metalliferous Deposits of Cornwall and Devon*.[25] Henwood gives descriptions of some of the lodes at Trewavas, though these were actually made in about 1838. South Lode was orientated 40° N of W and dipped between 80° SW and vertical; the maximum depth stated was 60 fathoms. The width ranged from 1 to 6 feet and seems to have averaged 2 to 3. The lode consisted of quartz with greater or lesser amounts of chlorite and altered granite carrying pyrite and chalcopyrite. Way Sowan Lode trended 25° N of W with a fairly constant dip of 70-80° N. It ranged in thickness from 6 inches to 4 feet, averaging about 3 feet and was worked to a depth of at least 68 fathoms. This was also composed of quartz and granite but mixed with tourmaline in places. Middle Lode trended 25° N of W and dipped 80° to the north. It was 2 feet wide and was examined at 43 fathoms from surface. North Lode had an identical orientation to Way Sowan Lode and a similar dip. Varying in width from 6 inches to 3 feet it averaged about 2 feet; at 68 fathoms (presumably from surface as there is no mention of the adit) it joined North Lode – this does not quite agree with the lode plan in figure 11 though this almost certainly dates from after the mine's closure. North Lode bifurcates just north-west of Old Engine Shaft where it passes through a crosscourse.

Although what actually occurred is uncertain, there is evidence that some of the original adventurers left the company and a new concern was formed around the end of 1839. This is revealed by a valuation of materials in September (which would be required on the relinquishment of shares), and a call for all demands for payment for materials supplied to be made to Collan Harvey by November, in order that accounts could be closed. Details of the valuation, made by Captain John Richards for James Plomer of Helston, have not been found, but a covering letter from Captain Joseph Vivian of Camborne stated the boiler of the great engine was in a very bad state.[26]

Apart from ticketings, nothing is heard of the mine until March 1840; at this time it was apparently doing well as a 1-128th share was advertised by Messrs Penaluna of Helston and Nettleton of Plymouth, for a minimum price of £250 (see figure 3, opposite).[27] On April 9th the mine had 253 tons of copper ore for sale, the result of two months' work.[28] Later in the month the mine had 87 tons of rich copper ore 'broken, raised, dressed, and sampled, on Tuesday last, in six days' 'under the spirited and good management of Capt. Wm. Martin'.[29] This production would not touch any of the 'recent good discoveries', though these were not described.

The first 'proper' mining report did not appear in the *Mining Journal* until May 1840, albeit from an anonymous source.[30] Dated April 20th, it stated that in Kemp's winze the lode was two feet wide, carrying good stones of ore. In Giddly's winze it was two to three feet wide, 'rich throughout'. The lode was less rich in the 56 fathom level but improving, while in the 45 it was three feet wide, two feet of which was 'a very rich lode of ore'. The lode in a rise above the 45 was also very rich. The lode had not been seen in the 32 in the previous 20 fathoms of driving, hence a crosscut south would be tried.

Two crosscuts were being driven north to the south lode and these were expected to connect by the end of the month; this does not make sense as there were no workings south of the lode to drive north from, unless it refers to driving north along one of the crosscourses inland from the cliff. The ground was described as 'favourable'. In the shallow adit the lode was 12 inches wide, in whole ground, *i.e.* with no other workings on it. This lode could be worked for a quarter of the cost of working beneath the sea. The Way Sowan Lode, at the 22 fathom level east of the new shaft, was two feet wide, of a 'kindly appearance'. There was no doubt that there would be a good discovery at the intersection of this lode with the Nimble Cutter Lode. The tribute pitches were said to be 'as good as can be expected for old ground', suggesting that that part of the mine had been worked for some time. The piece ended with an admonition from the Editor of the *Mining Journal* that the 'absence of the agent's name attaches an irresponsibility to the document itself '.

By this time operations were being carried out to some extent on five lodes: North, Middle, South (also called Old Lode), Way Sowan and Nimble Cutter. By May 1840 a new shaft had been sunk on the Way Sowan Lode.

In the autumn of 1840 Captain Henry Francis of Goldsithney visited the

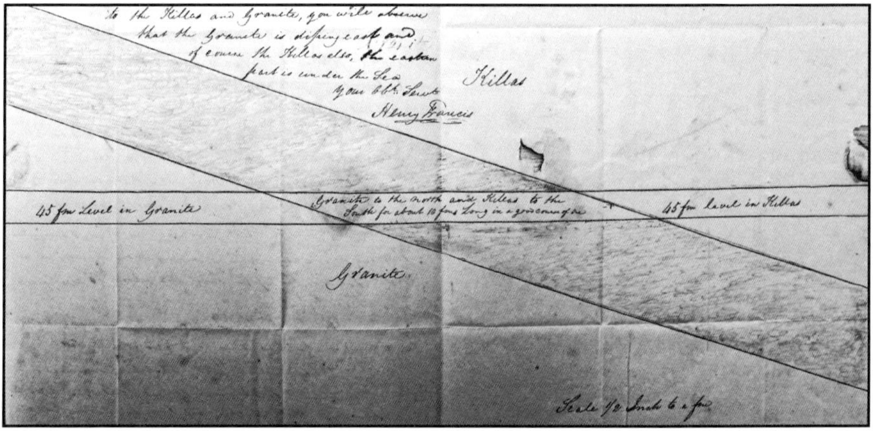


**Figure 4. Sketch of the granite-killas contact at the 45 fathom level by Henry Francis.**

mine on a number of occasions. Based on his visits and information provided by the manager, Captain William Martin, and another agent, he wrote several letters to John Hawkins FRS, describing geological features relating to the granite-killas contact seen in driving the 45 fathom level on the south lode. Rich ore had been passed through around the contact, but farther east in the killas, the lode was split and poor. A rise put up 6 fathoms above the level, in killas many fathoms from the granite, had again got into good ore. Francis finished his last letter, 'They have a north lode which is poor so far as it is worked on but they are not got forth to the killas in any of their levels as yet, and the lode they are raising the ore from being the South lode, does not make any ore as it gets into land'.[31] Prospects inland, it seems, were not promising.

During the early 1840s, possibly even the late 1830s, the mine was inspected by William Jory Henwood. Henwood, a geologist, was then retired and researching geology, then still a young and evolving subject, with regard to Cornish mines. Continuing work which was started by John Forbes some 20 years earlier, Henwood was at this time gathering information on temperatures in mines, although unlike Forbes he does not seem to have developed any theories.* From streams flowing from and near the Way

***John Forbes was a surgeon and the Secretary of the Royal Geological Society of Cornwall. After measuring the temperature of the air and water, and subtracting the heat he considered to have been caused by mining activity (candles, respiration, etc) he concluded that the earth became hotter**

Sowan Lode at the 43 fathom level, he recorded temperatures of 61°F. A stream of salt water at the east end of South Lode gave 57°, while water in the adit, drawn from 60 fathoms, was 56°.[32]

A letter to the *Mining Journal* in October 1840 from a shareholder expressed some surprise that Wheal Trewavas did not appear in the share list 'although I well know she is beginning to make a stir in the county, and likely to become very notorious'.[33] The reply from the *Journal* stated that the paper would be happy to print any reports sent from the mine, so long as they were authenticated by the captain's name. Following this the mine would doubtless appear in the share list. Encouragement for an increase in share prices came from increased copper ore sales, the total for the year being 1854 tons, worth £11,518 1s 3d.

Despite the criticism of the state of the boiler in 1839 it was not until January 1841 that the mine advertised for 'a good BOILER, from 10 to 13 tons'.[34] In 1841 copper ore sales showed a further increase to about £1,300 a month, and were maintained at this level for the next two years. As was normal in Cornwall the ore was sold at the regular ticketings, so that any of the South Wales copper smelters could have purchased quantities of the mine's production each month. The ore was sold as it lay on the mine surface, after dressing, the purchaser then having to arrange carriage to a port, storage at that port and shipment to South Wales. Pascoe Grenfell & Sons shipped their proportion of the ore from Porthleven, as probably did most of the other smelters, but sometimes a limiting factor was storage facilities at the quays, so some ore may have been shipped from elsewhere. Payment for the ore was made to William Vice of Truro and Charles W. Turner on behalf of the adventurers.[35]

Some technical information is available for the 45-inch engine for May and June 1841 in Lean's *Engine Reporter and Advertiser*.[36] The engine was single-acting and did a duty of 30,980,851 with an average of 6.74 strokes per minute; this was the equivalent of 413.1 gallons per minute, pumped by nine lifts. Two balance bobs were in the engine shaft, with an angle bob connected to 95 fathoms of fla-rods at surface. This indicates that it was pumping from

**with depth. This was an unusual theory for the early 19th century, particularly as there were no models or ideas of the earth's interior. Forbes later became court physician to Prince Albert and the royal household, was knighted and appointed a Fellow of the Royal Society.**

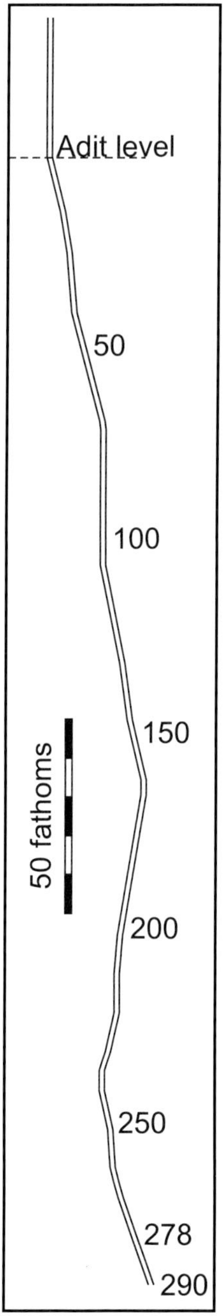


**Figure 5. Profile of Skip Shaft at Levant Mine, a good example of an underlay shaft. Despite its shape it was used for hoisting kibbles, then skips, from about 1840 until the closure of the mine in 1932. When re-used by Geevor Tin Mines from the 1960s cages for man-riding were installed. The ultimate hoisting depth was 278 fathoms, 1668 feet or 556 metres.**

at least one other shaft, presumably Old Engine Shaft to the west. If this is the case, the flat-rods would have run at least partly across the small cove immediately to the west of the engine house. Three bobs were connected to 85 fathoms of horizontal rods underground, with 36 fathoms of 'dry rods' in the shaft (*i.e.* the water in the shaft was 36 fathoms below the beam, below which, of course, the rods were wet). Steam for the engine was raised in three boilers, presumably two working at one time with one spare.

The stroke of the single-acting engine was 9 feet indoors but a curious, possibly unique, mixture of 4, 7 and 8 feet in the pumps (*i.e.* the outdoor stroke). This rather uncommon arrangement employed horizontal or inclined rods, moved by angle or V bobs, to alter the length of the stroke at each pump. The engine was pumping from two shafts 'diagonally', meaning that the shafts were both inclined to some degree.

In Cornwall at this time there were few, if any, truly vertical shafts. Cornish shafts were referred to as underlay or underlie shafts, the shaft being sunk on the course of the lode. These lodes were created by hydraulic fracturing (where the rock is fractured by the excess pressure of the mineralising fluids percolating through the rock), a process which does not lend itself to the words 'straight' or 'perpendicular'. There were also two benefits to sinking a shaft on a lode; one is that the rock adjacent to the lode tends to be a bit softer than elsewhere, the second is that mineral can be recovered whilst sinking, which can help pay for the sinking. The great disadvantage is that the shaft will become crooked, requiring extra equipment within the

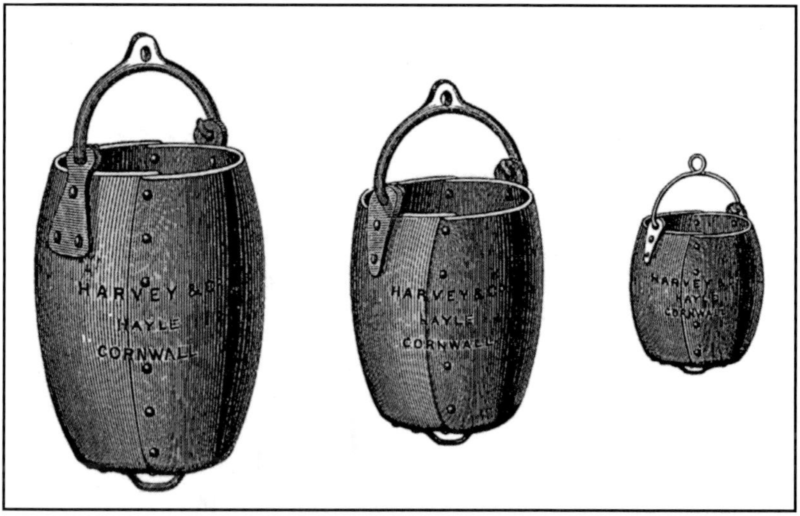


**Figure 6. Kibbles. Left-right: steam whim, horse whim and winze, from the Harvey & Co. catalogue. These were made from wrought-iron and over time wore out from the constant dragging along the sides of the shaft. A large mine could replace hundreds of kibbles in a year.**

shaft (angle, V or fend-off bobs) to help with the pumping.

In the same *Engine Reporter* the 'East Whim' (to separate it from a west whim?) was said to be hoisting kibbles weighing an average of 500 lbs, that the boilers were leaky and that the average duty of the engine was 3.3 million. Sadly, for a report on engines, this gives no clue as to the size of the cylinder. The engineer for both engines, probably consulting, was T. Tippet. The average duty of the 36 engines reported on was 53 million, so the two Trewavas engines were well below that figure.

An even greater surprise, if not actual shock, was experienced on 17 February 1842, when an earthquake affected the mine.[37] The quake, which happened at 8.30am, affected an area from the south coast near Helston to the north coast north of Redruth, covering an area bounded by Falmouth, Flushing, Penryn, Gwennap and Redruth. This most likely resulted from movements along the Start-Dodman-Lizard Thrust, a still active fault which extends along the south coast from mid-Devon past Lands End. While these shocks are rare, they continue to the present day.

At Trewavas, the miners in the bottom levels of the mine, and farthest out

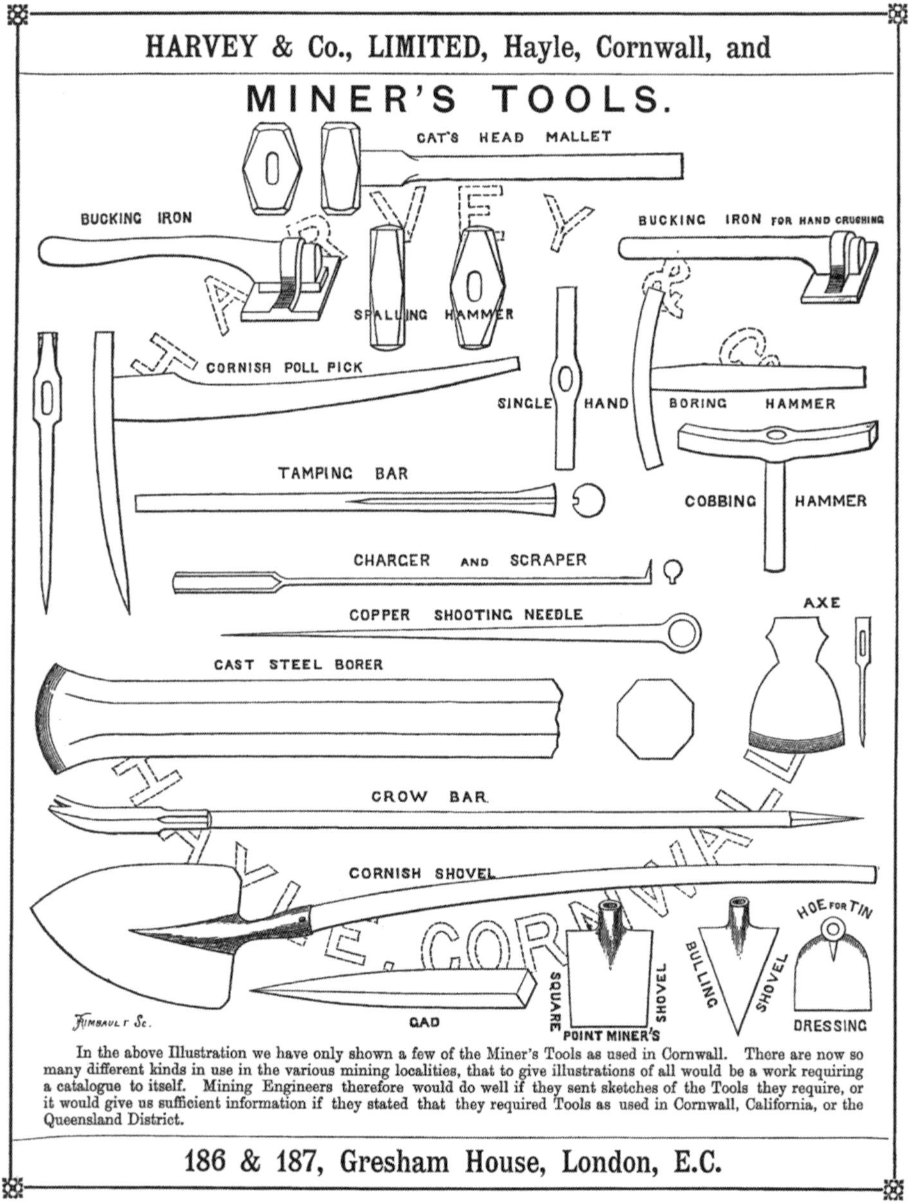


**Figure 7. A page from the catalogue of Harvey & Co. of Hayle showing various mining tools as well as some copper dressing hammers and mallets.**

to sea, rushed to surface, fearing that the sea had broken into the mine. Not having experienced an earthquake before they can be forgiven for confusing the sounds and shocks that they felt with rockfalls (or worse) within the mine.

The fact that the levels in the mine had extended some distance beyond the cliffs meant that working conditions would deteriorate because of increasing temperature and poor ventilation. Active ventilation beyond the use of manually operated fans, usually worked by a young boy, was unknown at this time. Hence the miners would be working in areas of great heat with poor quality air, particularly after blasting. The exhausting manual labour of the miner under these conditions was one of the major causes of premature death.

To remove ore, to raise over a level, to sink shafts and winzes and to drive levels meant breaking ground. Around the lode this could sometimes be done with a pick, but mostly it required the drilling of numerous shot holes. This drilling was carried out manually by miners using hammers and chisels, the latter known as 'boryers'. Drilling a single shot hole could take a day in certain rock types. Once the required number of holes had been drilled they were filled with black powder, carefully tamped with clay, fuzes added, and the holes fired. Death or injury could come abruptly from poor tamping, if a spark were struck, and copper tamping bars were commonly issued to miners. Safety fuse, invented by William Bickford in 1835, certainly reduced the number of deaths and injuries from blasting, but did not stop them completely. Prior to this either a train of fine powder or interconnected goose quills filled with fine powder formed the fuse.

A common practice for stoping, the removal of ore, was that of 'desuing'. Here the footwall (underside) of the lode was removed for a certain length and all of the waste rock removed from the level. After this the lode, or possibly just the mineralised part of the lode, was taken down. This method ensured that mixing ore with waste (known as 'dilution') was kept to a minimum.

Rock was broken down with sledge hammers and trammed away by young miners using narrow Cornish one wheel barrows. Often these youths would have rags wound round their knuckles to prevent injury from scraping the sides of the narrow levels. The barrows of ore were eventually tipped into kibbles – rounded wrought iron buckets – and taken to surface by the hoisting engine if present, or horse- or man-power.

Figure 8. 'Bal maidens' spalling copper ore.

At surface, the copper ore was usually dealt with by a small army of women and girls, with a few men, breaking the ore with a variety of hammers to picking out the waste to produce a concentrate. The end result of this were pieces of ore about the size of a walnut, free from any waste rock.

At the beginning of March 1842 a 16 year old called Henry Richards had been wheeling a barrow over a board across the shaft when it gave way. The unfortunate young man then fell from the 22 fathom level to the 56 fathom level (34 fathoms or 204 feet) and was killed instantly.[38]

Early in November the middle boiler of one of the engines burst 'with a tremendous crash', doubtless causing consternation to all those who heard it.[39] The accident was fortuitous as several miners, along with the engineer, had not long left the 'engine room', presumably the boiler house, though no casualties were reported. As several articles of clothing were lost it would seem that either the house or boiler house was being used as a changing room.

The accounts, shown to the shareholders in December 1842, were sufficient to allow a dividend of £5 per share to be given.[40] On 31st March 1843 a

Figure 9. Cobbing copper ore from a hand-barrow.

**Figure 10. The pictures opposite show rather delicate, nicely dressed, women at work on the copper floors. This image, of three bal maidens filling a wagon at Carn Brea Mine, comes from *Harper's Magazine* of 1881 and much more accurately depicts the back-breaking nature of the work.**

third accident was recorded on the mine.[41] Two men, Richards and Truscott, were in the process of tamping a hole when it suddenly exploded; Richards was killed and Truscott severely wounded (it is possible that Richards was related to the youth who was killed in 1837).

In July that year the value of the mine's shares suddenly dropped from 125s each to 90s. This lack of faith in the mine was probably a reaction to the reduction in ore production which fell from just over 2,300 tons in 1843 to 1,600 tons the following year; with the exception of a three week period in March 1844 the value of the shares never again rose above 100s.

In October 1843 Richard Gilbert, one of the miners, was charged with 'having feloniously taken, removed, or concealed, a quantity of copper ore, on the 18th Sept., in Trewavas mine, in the parish of Breage, with intent to defraud the adventurers'.[42] The prosecution was conducted by Messrs Shilson and Plomer, the latter a Helston solicitor and the mine's purser, while the defence was carried out by a Mr John.

Captain Michael Williams Martin showed that Gilbert had taken a pitch on tribute at 6s in the £1 for September. Two lodes were then being worked, the highest tribute being 10s and the lowest 1s. These prices suggest that the lodes were between comparatively and fabulously rich, although this does not mean that any of the other lodes were. Captain Henry Martin oversaw the dressing floors and was present on September 18th. There had been about 10cwt of adventurers' ore on the floor which Martin had ordered to be taken to the higher end of the bucking house. Some tributers' ore, part of which belonged to Gilbert, was also there and was also moved into the bucking house.

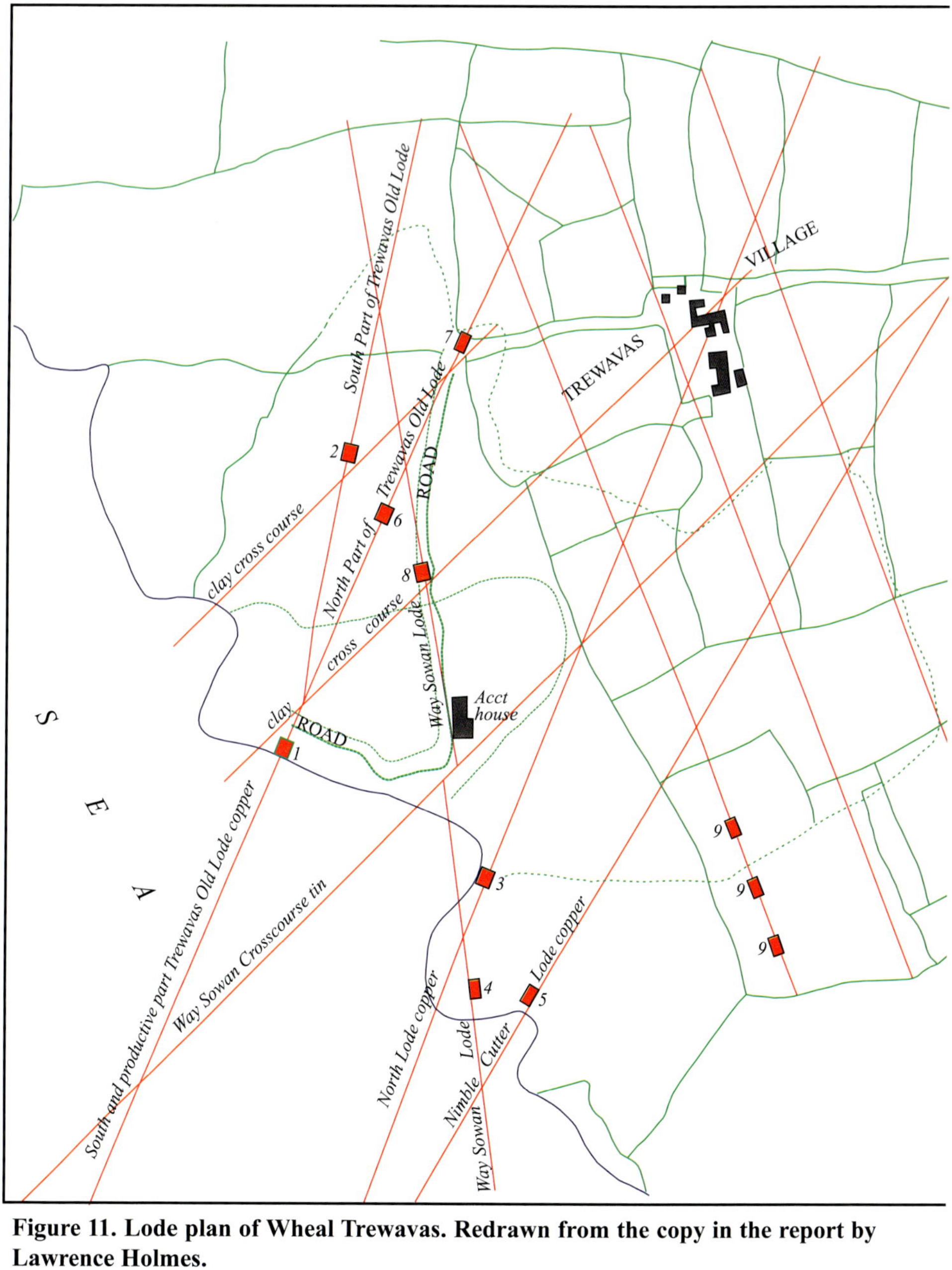


**Figure 11. Lode plan of Wheal Trewavas. Redrawn from the copy in the report by Lawrence Holmes.**

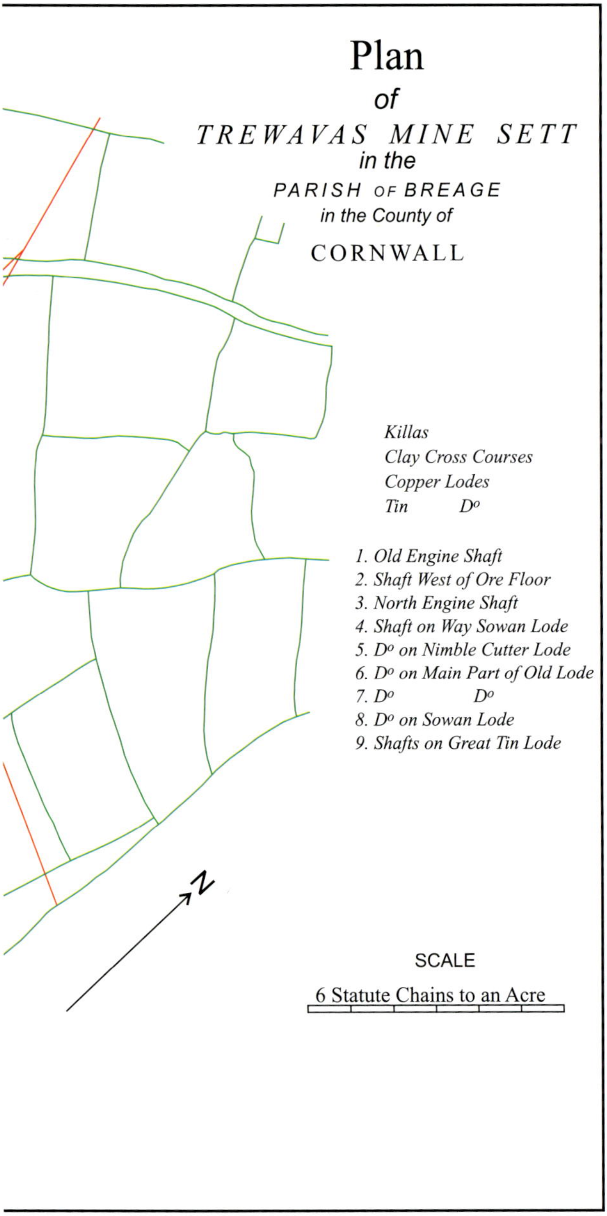


After visiting another part of the mine, Captain Martin returned to the bucking house, only to find that the owners' ore had been taken away. Gilbert was in the house with the bucking girls. On being asked who had scraped away the ore, Gilbert replied that he had but had not taken any; if Captain Martin thought any was missing he could take it from Gilbert's pile to make up any difference. There was apparently not much difference in the value of the adventurers' and tributers' ore.

On cross-examination, Captain Martin said there was no partition between the ore piles on the shelf, but there were for ore that had been bucked; these were 10 inches apart (though there seems to have been no physical partitions). Two women were employed to buck the adventurers' ore and six for the tributers' ore. Several of the women had given evidence that Gilbert had shovelled down some of the owners' ore and had been told not to do so. Gilbert had then told them that he would take any blame. The women had also stated that the last of the tributers' piles was close to that of the owners', although there was no difference in their appearance.

Captain Martin had also admitted that he only knew the owners' ore when it arrived on surface by the tokens that came up with it (each pare, or group, of men stoping would have a different token; by this means the amount raised could be tallied so that the men could be paid). Mr John's defence was therefore that there was no actual proof that the pile called the adventurers' ore was really so, and it was therefore not possible to prove any felony. Following this, three 'respectable witnesses' gave evidence for Gilbert's 'good character and honesty'. This seems to have worked and Gilbert was found not guilty.

It was not common for a case like this to reach the courts and one wonders how Gilbert, presumably not a wealthy man by anyone's standards, could have afforded legal representation. However, having won the case, it seems unlikely that he was able to continue his employment at the mine.

A snapshot of the mine for 1843 is given by Joseph Yelloly Watson.[43] Four or five lodes were in the sett, lying close together, the deepest level being the 70 fathom level. From 1835 to October 1842 the mine had returned 10,838 tons of copper ore for £71,981 13s, from which had given a profit of £7,680. About 200 people were employed and the mine was now making £4,000 profit per year.

1844 saw a decline in production; evidently the richest ore ground in the levels opened on South Lode was becoming exhausted. To sink below these levels was probably beyond the capacity of the existing pumping arrangements. To overcome this a 70-inch engine would be erected on the Old Engine Shaft, and also, as the ore pitched out under the sea, the diagonal shaft was used from about this time to reduce the cost of drawing ore to surface, improve ventilation and facilitate further development. This shaft is shown in figure 12, extending southwards from New Engine Shaft. This shaft is confusing; the first mention of it dates from about the same time as the erection of the 70-inch engine and one would expect that this engine would have been put up near the shaft's collar, inland and uphill from the engine house, the more efficiently to service it. However the engine house is on Old Engine Shaft on the cliff edge and figure 12 does not show the shaft crossing Old Engine Shaft. It is possible that the shaft dates from an earlier period of the mine's history, at least in its upper part, and that the lower part was deepened as the workings extended out to sea.

Diagonal or inclined shafts had been used a number of times in the south west, for example at Wheal Friendship and Wheal Crebor in Devon; one had been started at Tamar Silver Lead Mines towards the end of 1843 to run out under the estuary of the River Tamar where most of the ore in that mine lay. That at Trewavas was perhaps the first planned to work ore out under the open sea. Other, better known, examples were subsequently sunk in the St. Just area at Botallack, Boswedden, Wheal Edward and Geevor. However a common theme regarding all these West Cornwall examples is that they were relatively short lived, the expected continuation of the ore in the lodes they worked proving to be either limited or non-existent, or the mines succumbing to economic pressures before sufficient development had been carried out to prove the ground offshore. Many other mines had inclined haulage shafts sunk down the dip of relatively flat dipping lodes, in contrast to the above cases where the inclines were constructed to follow the pitch of an ore body or expected ore body.

In February 1844, John Truran, one of the agents, was awarded a £5 prize for an essay on general mining.[44] Truran was later an agent at Ding Dong Mine and had a number of his geological essays printed in the *Mining Journa*l. His essay, produced under the heading Cornish Mining – The Wizard of the West, the *Journal*, rather tongue in cheek, referred to the 'lucubrations (a learned or pedantic piece of writing) of Capt. John Truran' followed by the rather wordy work:

> The mode under which bunches of ore occur, has its importance - but in the present state of mining a description is not easily made. Every one acquainted with meteorology must be familiar with the method adopted by the philosopher, in setting forth the different kinds of vapour that float or lie stagnant in the superior regions of the atmosphere-he uses the terms 'Cirrus-stratus Cumulus'. Now, were the whole lode laterally exposed to view, so that the mineral could be seen, it would be found that the last named terms are not inapplicable; but the prevailing form is cumulus, or rather cumulus and columnar combined, with more or less dip. This is important, as much mining capital had been wasted, by extending too many levels in a mine; when,

> if a miner has directed his operations to depth, more especially at the commencement, and then to pass a level through the lode, this would pass through every column or bunch of ore, thereby ascertaining, with the expense of one level, all that could be seen with many.

On May 6th the account for January and February was presented.[45] This showed:

| | | |
|---|---|---|
| To amount of cost for January and February | £2,793 15s 11d | |
| Copper ore sold | £1,899 15s 11d | |
| Less Lord's dues | 105 10s 10d | 105 10s 10d |
| | | 1,794 5s 1d |
| Received by Rev. C. Rogers, on account of sinking shaft | 200 0s 0d | 1,994 5s 1d |
| | Loss | 799 10s 10d |
| In debt end of December | | 1,741 1s 11d |
| Total debt | | 2,540 12s 9d |

This indicates that the mine was somewhat less successful than indicated by Joseph Watson, though, as a share dealer, he was almost obligated to make a mine look as good as he could have.

In June 1844 the mine's committee adverted the mining public to the fact that it would soon be offering for sale the 45-inch pumping engine and whim engine, along with sundry spare materials.[46] Unfortunately the mine was about to find itself the centre of the rumour of a disaster, fairly quickly denied but which seems to have sprung up again over a century after the event.

At the end of June 1844 the *Mining Journal* said that it was 'in our power to contradict a rumour in general circulation, which was calculated to have an injurious effect of the interests of adventurers in Trewavas Mine'.[47] A rumour had commenced in mining circles that the sea had broken in and the mine was flooded. John Plomer, the purser, had written to the *Journal* stating that he was 'just returned from the mine' and was ignorant of any circumstances that could have given rise to such reports. Plomer also wondered how such rumours could have started without any information to feed them.

A letter in the same edition of the *Journal* stated that the rumour prevalent in the city was that the sea had broken into the mine and that it would therefore be abandoned. The rumour had been traced to one individual, not named. The correspondent had heard from the mine that 20 fathoms of rock separated the mine from the sea and wondered about the motives of the person spreading the rumours.

The following week the *Journal* reproduced a letter from John Plomer asking for the name of the individual responsible for the false report.[48] At a meeting on the mine, on July 3rd, the following resolutions had been passed:

> That the purser be requested to ask favour of the Editor of the *Mining Journal* sending him the name of that person.
>
> That the report must be considered as originating solely in interested or malicious motives.
>
> That the purser return the thanks of the adventurers to the Editor of the *Mining Journal* for having so promptly made enquiries, and publicly contradicted the rumour.

The editor of the *Mining Journal* had replied that it was not a favour that was required, naming the culprit as Frederick Hill of Helston, writing to his brother in London; apparently the rumours had made some people part with shares for only a quarter of their market price.

A letter from one of the mine's (anonymous) shareholders appeared the following week:[49]

> Trewavas Mine – Charge and Explanation
>
> Sir – As an out adventurer, and, hence, not a subscriber to the very admirable resolutions passed at the mine on the 1st inst (and forwarded to you by our worthy purer, a printed copy of which I received, and which, had I been in town, should have sent you), I have to thank you, in common with my co-adventurers in Cornwall, for the exposure you made on the foul attack on our property. We now know our men, and I, for one, shall mark them well. I do not blame Mr Frederick Hill, of Helston – for although he is a lawyer, I believe him to be honest intentioned, and I have had some business with him, which has given me satisfaction; as regards his brother, Mr T. I. Hill, of Castle-court, in

the city, I can only suppose the letter was sent to him confidentially, and with the intention of putting his relative on guard – such a report, in all probability, having been promulgated in the county. How far the latter was warranted in giving it publicity, or how far you were justified, in one sense, in noticing it in your columns, I leave you two gentlemen to determine. But I for one thank you sincerely for your honesty, although I am not prepared, in the absence of further information, to admit you acted with your usual discretion. At the same time, allow me to add, I have no right to suppose you would ever act without being able to afford a sufficient explanation. My only object in writing is, to prevent confidential communications, where incautiously conveyed to others, being published even in your columns. – London, July 9.

An Out-Adventurer

In a most gentlemanly way the *Mining Journal* thought that both of the Hill brothers had in some way been deceived, though if either of them were actually the source of the misinformation it could not be proved. There must also be a small possibility that the story of the earthquake of February 1842 was recirculated in a garbled fashion and picked up by Hill. Hill's letter appeared in the same issue:

Trewavas Mine

Sir – I beg you to insert the following statement in your next publication:- In writing to my brother, I incidentally mentioned a report that was current in the neighbourhood, that the sea had broken into Trewavas Mine, and I did so on authority which I believed could be entirely relied on. I regret that I should have incautiously given credit to the report, and still more so that it should have affected the character or credit of the mine. It is due to myself to add, that the communication was intended only for my brother, though I have no reason to complain of his having mentioned my name as his authority; the circumstance has enabled the committee of the mine to refute an impression which generally prevailed in this neighbourhood. I have no interest involved, and sincerely hope this mine may long prosper, and benefit the very liberal lord and the adventurers, with many whom I have the pleasure of being on terms of personal intimacy. I have

> thought it right to see the purser, Mr Plomer, and furnish him with the names of two persons on whose authority I credited the report.
>
> Helston, July 8. Frederick Hill

Frederick Hill was, however, no stranger to dabbling with the truth. A letter in the *Mining Journal* in May 1836 referred to a previous letter signed by John Silvester and witnessed by Hill.[50] In this letter Silvester and Hill appear to have been operating a scam by calling themselves agents for the bounders of Duchy of Cornwall lands; Silvester had in fact been one such previously but had since been dismissed. Messrs Bartlett & Beddone, solicitors in London, were writing on behalf of the Duchy, stating:

> . . . Why, Mr John Silvester and Mr Frederick Hill know as well as any two men in Cornwall, that their vaunted bound claims are never to be maintained for a moment, unless they have continually worked, are continually paid, the duchy lessee his toll, and we give them and the county notice, that if they find any adventurers foolish enough in this enlightened day to take a sett in duchy land under a bounder, we will, immediately on tin being raised, bring an action against such party for the full amount of the customary dues, without sharing any part with the bounders, as in the day of 'perfect harmony' alluded to by Mr John Silvester 'the agent for the several tin bounders in the manor and parish aforesaid'. As to the threat in the advertisement of taking away the tin from certain mines, which are left to imagine themselves in the words 'et cetera' of Mr Hill and Mr Silvester 'the agents …'. go on any such expedition, we hope the adventurers in those mines know how to defend themselves from plunder by day or by night; and as to any legal measures to deprive them of their tin, anyone who knows anything of present mining rights in the county of Cornwall, will guarantee them against all the bounders for half a crown.

The 45-inch engine, along with assorted other pieces of equipment, including the pitwork and pumps, was eventually advertised late in July and August for sale on August 14th.[51] The engine was not sold and was advertised again in September (with no sale date);[52] October, for sale on the 25th[53] and November, for 9th December.[54] It is presumed that the engine was sold then as there appear to be no further references to it in the local newspapers.

A mining report appeared in August, dated July 10th and signed by captains W. Martin, W. M. Martin and J. Truran.[55] The lode in the 85 east was from two to three feet wide; a good course of ore. A raise had been put in the back of this level to connect with the diagonal shaft. A good course of ore could be seen in Bawden's winze, in the bottom of the 70. In the 56 and 66 levels the lode had been disordered by flooken (a fault, often containing clay) but was producing occasional bunches of ore. The tribute pitches were not looking very well and were not expected to until the winzes were connected to the 85 fathom level. More ore was expected to be met with in the 70 and 85 levels which the diagonal shaft would help access. A new engine (a 70-inch, with three boilers) had been acquired and was already in full operation. Another sale notice, in October, advertised not only the 45-inch engine but also a 16-inch whim.[56]

Yet another accident occurred in August.[57] Henry Pascoe, of Torleven (probably the farm which is now in the centre of Porthleven) had been in a shaft beneath a kibble when its rope broke. The kibble struck Pascoe on the head, killing him instantly; the kibble would have weighed several hundred pounds, and Pascoe would have had no chance. The *Mining Journal* stated that one of the strands of the rope had been previously broken, and that splicing it had been suggested.[58] On the previous day Thomas Magor, a well-known Cornish wrestler, had been killed at Wheal Vor, following an explosion whilst tamping a shot hole.

Nothing is heard of the mine until the following year. In late March it was reported that a lode had been discovered which was likely to prove valuable.[59] 'It has the promise of great richness, and will, it is expected, realise the sanguine expectations expressed by the mining captains'.

The report for 5th May 1845 stated that the diagonal shaft was completed to the 85 fathom level, the end being 116 fathoms from the adit, with a further 8 fathoms as a sump.[60] The end of the 85 was 48 fathoms from the shaft, containing a lode worth from £20 to £80 per fathom. Pellew's winze had connected to the 85 and was driving towards Bawden's winze, which was sunk 13 fathoms below the 70. Dunn's winze was 102 fathoms east of the diagonal shaft and 54 fathoms east of the end of the 85. The level had contained good ore but the last 4-5 fathoms had been driven through a flooken. The accounts for January and February showed:

| | | |
|---|---|---|
| Amount of cost for January and February | £1,781 14s 1d | |
| Copper ores sold January and February | £820 17s 5d | |
| Less lords' dues 45 12s 1d | 775 5s 4d | |
| Loss | | 1,006 8s 9d |
| Book debt end of December 1844 | | 2,569 4s 0d |
| Total deficit | | 3575 12s 9d |

This shows the deficit having increased by around £1,000 (60%). Bearing this in mind one wonders why the management was expanding the equipment on the mine, presumably as a basis for deepening it. In fact this sort of activity in Cornish mines was very typical, throwing good money after bad and taking optimism far beyond any reasonable bounds.

In September 1845 the mine advertised for a principal underground agent.[61] It is unclear if this was an addition to the staff or a replacement; the man was expected to be 'thoroughly conversant with his calling'. By 6th October the sumpmen were cutting a plat in the diagonal shaft at the 85 fathom level.[62] The shaft had resumed sinking and was now 8 fathoms below; the lode was poor and the ground hard, but improvements were anticipated. Pellew's winze was eleven fathoms below the 85; no lode had been taken down, or would be, until it was deep enough for another level. In Christopher's winze the lode was worth about £20 per fathom; it was 2½ fathoms below the 86 (presumably the 85) fathom level. The end of the 85 fathom level was now over 60 fathoms east of the diagonal shaft, in which distance the lode varied in value from £10 to £25 per fathom; over the last few fathoms it had been less productive than desired though it was once again in soft ground and in a rich course of ore.

To connect this winze with the eighty-five fathom level there were about sixteen fathoms to drive. East of the winze the lode had passed through good bunches of ore, but at present it was poor and divided. In the thirty-two fathom level west the miners were driving on the north part of the lode, which was poor, the ground hard.

The accounts for 11th November 1845 were for July and August.[63] Costs and merchants' bills amounted to £1,559 2s, while ore sold minus lord's dues totalled £1,047 9s 8d; the loss on two months was therefore £511 16s

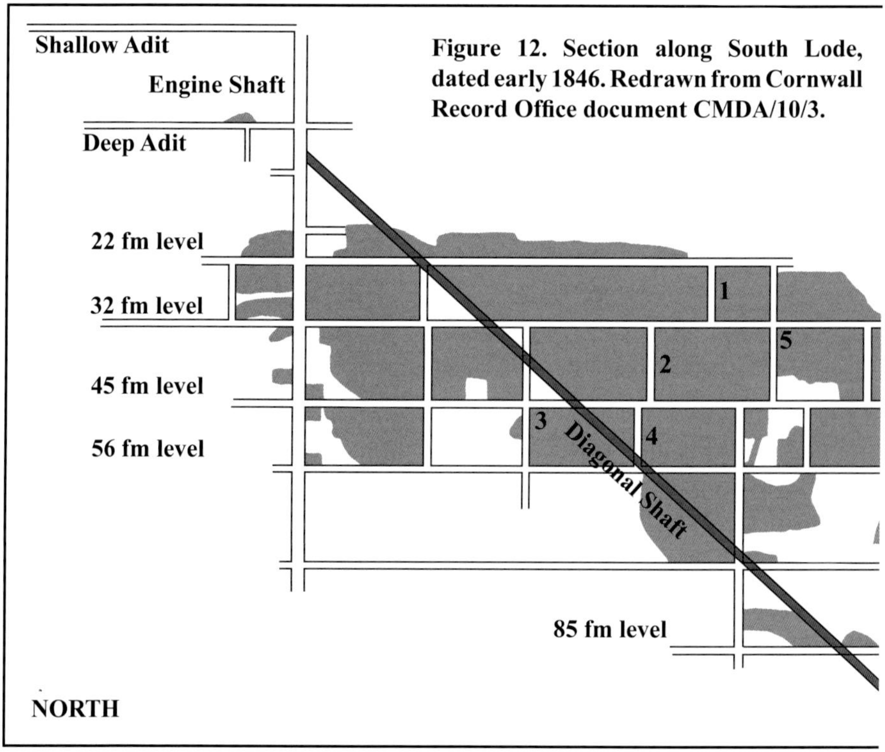


**Figure 12. Section along South Lode, dated early 1846. Redrawn from Cornwall Record Office document CMDA/10/3.**

6d. The cost book debt to the end of June was £4,103 11s; minus a call in September, amounting to £1,280, the total debt was now £3,335 7s 4d. The debt on the mine was now increasing and many calls remained outstanding; the purser was directed to proceed against those in arrears after fourteen days of the meeting. These calls must have been onerous; the September call was £10 per share, a large sum. Another call, of 10s per share, was made, payable on December 1st. Underground it was decided to abandon the 32 fathom level (possibly it was approaching the sea bed) and to lay a 'railway' (actually a tramway) in the diagonal shaft and flat-rods in the 45. The latter was presumably to enable pumping from one of the winzes.

A few days later the mine advertised for sale a 10-inch pressure engine, 'six feet stroke, with brasses all complete'.[64] The notice included various bits of pitwork (pump rods) which were worked by the engine. This indicates that the engine was used for pumping, rather than hoisting. It is not clear which shaft it worked on, though Old Engine Shaft would be an obvious choice as

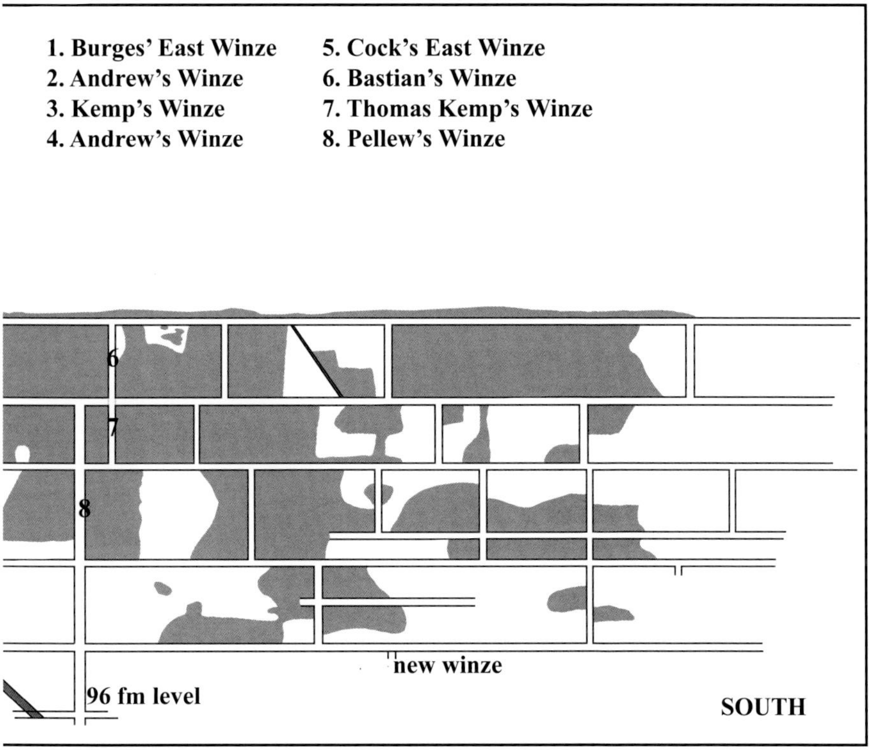


the engine would require a head of water to work it.

The report for the following week stated that the lode in the 85 fathom level was from two to three feet wide, in soft ground with small bunches of ore.[65] Pellew's winze was now eleven fathoms below the 85; the new, 96 fathom, level had been driven three fathoms west and within two fathoms of the line of the diagonal shaft; this had another five fathoms to sink to meet the level. By the end of the month it was hoped to have pitwork fixed in Pellew's winze to drain the 96; possibly this may have been connected to the flat-rods in the 45. This, in addition to laying the tram road in the diagonal shaft, was estimated to save the mine £60 per month on an outlay of £400.

The lode in the 32 fathom level, west of what was now called Roger's Engine Shaft, was small and poor, in hard ground. To work this part of the sett, 'on the less expensive and more effectual method', it was recommended to extend flat-rods 110 fathoms west from the engine at surface at the surface to a shaft sunk about twenty fathoms from surface, on the old or

South Lode. By sinking the shaft and extending a level east and west the lode could be developed in much less time, and for less money, than waiting for the extension of the 32 from the main part of the mine. In addition, the water pumped from here would be fresh and could be used for boilers and condensers, which would save the mine about £200 a year.

Finances had improved very slightly by January 1846.[66] The accounts for September and October being:

| | | |
|---|---|---|
| Labour cost | £992 17s 6d | |
| Merchants' bills | 641 2s 1d | |
| | | £1,633 19s 7d |
| Ores sold (less dues) | 627 3s 1d | |
| Sale of materials | 860 5s 6d | |
| | | 779 14s 1d |
| Book debt to end of August | 3,335 7s 4d | |
| | | 4,109 1s 5d |
| Call of 10/ per share | 128 0s 0d | |
| | | 2,829 1s 5d |

Despite this, however, the value of the mine's shares dropped from 40s each to a mere 10s each at the end of January.[67] The shares retained this value until March, when they suddenly rose to 32s 6d, then dropped to 30s in April, following which they were not quoted by the *Mining Journal*.

It is from about this time that the only underground plan dates, a section (or vertical longitudinal projection), along what is thought to be South Lode, the mine's most productive lode (figure 12).[68] The grey areas are stopes, where the ore had been removed. This is interesting as it shows that the area of copper mineralisation does not dip, as might be expected. Unfortunately there appears to be little mineralisation at depth and to the south; with the dip of the sea bed this meant that ore reserves were limited both horizontally and vertically. Some of the winzes, underground shafts connecting not more than a few levels, are shown here. This drawing also helps to appreciate the difficulties in ventilating the works, particularly as it was mostly done passively; single-shaft operations such as this were notoriously difficult to ventilate and workings conditions would have been very poor.

**Figure 13. Looking north-west to the mine from Poldhue on The Lizard: Old Engine Shaft left, New Engine Shaft right. Photo: Author.**

**Figure 14. Looking west from the head of the inclined mine trackway past the New Engine shaft chimney and engine house towards Old Engine Shaft. Photo: Adam Sharpe, Cornwall Archaeological Unit.**

**Figure 15. Engine house on New Engine Shaft. In the background some aplite/ pegmatite sheets can be seen traversing the killas at the Tremearne Cove SSSI. Photo: Ainsley Cocks, World Heritage Site Office.**

**Figure 16. General view from the east. Photo: Adam Sharpe, Cornwall Archaeological Unit.**

**Figure 17. Engine house on Old Engine Shaft with shaft capstan at rear. Photo: Adam Sharpe, Cornwall Archaeological Unit.**

**Figure 18. The shaft capstan plat at Old Engine Shaft. Photo: Adam Sharpe, Cornwall Archaeological Unit.**

**Figure 19. Aerial view of the mine from the south. Tregonning Hill is in the**

**background. Photo: Barry Gamble.**

**Figure 20. Fine waste on the site of the copper dressing floors. Photo: Adam Sharpe, Cornwall Archaeological Unit.**

**Figure 21. The rear wall of the 70-inch engine house showing the precarious state of the masonry and remaining lintel. Photo: Adam Sharpe, Cornwall Archaeological Unit**

**Figure 22. Masons at work on the top of the east wing wall of the 70-inch engine house. Photo: Adam Sharpe, Cornwall Archaeological Unit.**

**Figure 23. The pumping engine house on Old Engine Shaft with its cover of scaffolding. Photo: Adam Sharpe, Cornwall Archaeological Unit.**

**Figure 24. Looking down from the top of the bob wall at the Old Engine Shaft, right, and whim shaft, left. Photo: Ainsley Cocks, World Heritage Site Office.**

**Figure 25. The scaffolding route down to the engine house on Old Engine Shaft. Photo: Ainsley Cocks, World Heritage Site Office.**

More trouble for the mine appeared in the form of a letter which had been received from Edmund Turner, a partner in the mine's bank in Truro (and a member of the mine's committee), in December.[69] A meeting to discuss this was held in Helston on 13th January 1846. The letter drew the adventurers' attention to the state of their account, which currently had a negative balance of over £5,000. Turner and his partners required a 'considerable reduction therein'; in this case an immediate reduction of at least £3,000 (the *Penzance Gazette* of 21 January 1846 says £2,000). It was therefore resolved to hold a special meeting at Pearce's Hotel, Truro, on the 21st, 'for the purpose of complying with file requisition of Messrs Ricketts, Enthoven, &c., and of dividing the balance shown in the last accounts'.

Ricketts, Enthoven & Co. was based at 8 Moorgate St. London, the company consisting of Frederick Ricketts, merchant, Henry John Enthoven, merchant, Edmund Turner, merchant, James Oliver Mason and Trevenen James, who joined in a partnership opening banking houses at Truro, Penzance, Falmouth and St. Columb.[70] They did this by acquiring the banking establishments of the Western District Banking Co. at Truro, Penzance, Falmouth and St. Columb and opening business at those places in September 1844. On 25 April 1846 Edmund Turner and James Oliver Mason retired and the firm became Ricketts, Enthoven & James. On 26 January 1847 Ricketts and James were in commercial difficulties, their chief creditor being The Governor and Company of Copper Mines in England. The partnership was dissolved on 28 April 1848 and the four banks were wound up.

It is possible that the problems that the Trewavas management experienced with Ricketts, Enthoven & Co. was related to the company's own financial problems.

In reporting the division of the loss between the adventurers at a special meeting held in January 1846 the *Mining Journal* remarked that the calls in arrears amounted to nearly £1,800 and that until eighteen months or so previously bimonthly dividends of £5 per share were being made, while three years ago the shares stood at ten times their present level.[71] If its information was correct, Edmund Turner, while acting as a member of the committee, had made advances of the bank's money for the purpose of doling out dividends, and was at the same time 'a seller of shares at the high prices maintained by this chicanery'. Turner, while claiming to be the largest shareholder, refused to reply to the charge unless the name of his accuser was made known to him.

Then a Truro correspondent reported that it had been discovered that the last dividend paid by the company came from money borrowed from the bank. This indicated just how little scrutiny was paid to the committee, though the shareholders were doubtless more than happy to keep receiving money, regardless of its origin.

To rub salt into the wound, Henry English, the editor of the *Mining Journal*, also decided to join the fray with an article entitled *Publicity as Associated with the Cost-Book System.*[72]

> Trewavas Mine. — The state of this mine, from the accounts received, would appear to be of a most unsatisfactory nature, inasmuch, not to advert to the price at which the shares may be obtained, being about one-tenth of that realised some three years since, a debt of 6,000/, or thereabouts, has been contracted with the bankers at Truro. It is hardly necessary to observe, that on the number of shares, into which the mine is divided (128), that the debt owing alone to the bank, exclusive of any merchants' accounts, or other liabilities, is equal to 40/ per shares — a matter which, doubtless, has its effect, as such must be added thereto, as the virtual cost of the share. It is true, some 1,200/ or 1,400/ may be due as arrears on calls already made, and which, if paid up, would reduce the balance so far. While, if we are rightly informed, certain measures, if not pending, are, at least, contemplated by the bank in question, to enforce the payment of the balance, and which thus renders the matter somewhat serious. If we recollect aright, the dividend payable every two months, up to a period of some 18 months since, was 5/ per share, or 39/ per annum, while the present price of the share is not equal to one years' dividend. We have been given to understand, that Mr E. Turner, MP, is a partner in the bank alluded to, and one of the managing committee; or, at last, was so at the time of dividends being paid, and, as we are told, loans being obtained from the bank, in which he is interested, such being necessary to effect the object. If, again, our information is correct, this gentleman, while he doled out the dividends, and, at the same time, made advances from the bank, was a seller of shares at the high prices maintained by this system of chicanery, and, therefore, might be supposed to be interested in pursuing such a course. We should hope such was not the case, but

the rumour having reached us, we can only say our columns are freely open to any reply to the charge; while, under any circumstances, it is to be hoped that Mr. Turner, as associated with the mine, will endeavour to prevent such proceedings on the part of the bankers, as are calculated to militate against the interest of the adventurers, and, more especially, those who have been innocent of the system observed, and which, if our information be correct, cannot be too strongly deprecated.

A reply from an unlooked-for source appeared the following week:[73]

WHEAL TREWAVAS MINE

Sir, — I had no opportunity of looking into your Journal of the 17th inst, until my arrival in London, this morning. What you state, under the head 'Wheal Trewavas,' is perfectly true; on the other band, your paragraph, under the title 'Trewavas Mine,' is full of untruths. But, unless you think it right to give me — privately, or otherwise — the name of the author of the 'rumour' — which your paper has given currency to — I do not feel bound to correct or reply to it. I wish, however, to state, that I am the largest shareholder in the mine, and that at no time have the adventurers in it been threatened by the bank with proceedings at law, nor are such even now contemplated.

Believing, as I do, that you desire to serve the mining interest of Cornwall, you will, perhaps, permit me to suggest to you, that the insertion of such articles, on questionable authority, as your Journal has lately contained — attacking those persons who have invested their surplus capital in mines — is not calculated to promote it.

It may be your opinion, that a reasonable advance made by bankers is undesirable in mining. I can only tell you, that the most profitable copper mine in Cornwall was re-worked by an advance of money made by the bank of that day, and in which I was also a partner.

Ed. Turner

London, Jan. 22.

Henry English was happy to include the letter, but noted that:

As, however, a meeting was held on the 21st, when this subject would

> come under the notice of the adventurers, we are glad to find our wishes thus anticipated. The rumours to which we adverted, were generally promulgated in town, by parties interested in the mine, who will, doubtless, observe on the letter of our correspondent. As regards the opinion we may entertain on the subject of advances by bankers, this is a question to which we shall return.

The meeting of shareholders at Pearce's Hotel in Truro was reported in February.[74] At the meeting it was resolved that

> That inasmuch as the practice of charging interest and commission half-yearly on a bank account is unusual among bankers in this county, the excess of interest arising from such charge, in the account between Messrs. Ricketts and Co. and this mine, be disallowed.

In effect the mine was telling the bankers its interest charges were unfair. Messrs. Willyams and Tweedy, the former on behalf of the Miners' Bank, and the latter on behalf of the Cornish Bank, had offered to make any advance for the general working of the mine, in equal proportions, in case the necessity should arise. Thanks of the meeting were given to Willyams and Co. and Tweedy and Co., for their kind offers. The purser was authorised to avail himself of the offer, 'if he shall require it for the use of the mine, to the extent of not exceeding 1,000/'. A letter from Canon Rogers was also read, offering to give up half the dues already payable, on condition that the working of the mine be prosecuted. It was also stated that a new system of management be adopted on the mine and immediate measures were to be taken 'for the appointment of persons to whom shall be intrusted the conduct of the mine'.

Changes were to take place immediately. The thanks of the adventurers were given to the present management committee for their past services, but their services were to be dispensed with for the future. J. G. Plomer was to be appointed purser on a salary of six guineas per month and William Davey to be storekeeper and clerk for four guineas per month. Two permanent agents, 'day and night captains', were to be appointed as soon as possible, and Captain Stevens was appointed one of these, the other to be appointed at the meeting on March 2nd (the two vacancies were advertised prior to the meeting[75]).

The purser was to give notice to such agents who were not to be reappointed, their jobs to terminate at the end of February. Mr. Reginald Rogers gave notice that he would recommend the appointment of Captain Michael W. Martyn at the next account, on his present salary as superintending agent.

Another letter regarding the situation appeared in the same edition of the *Journal.*

> Sir. – In your Journal of Saturday last, appears one or two letters on the mismanagement of the above mine, and a notice of a meeting, convened to consider a proposition by Mr. Willyams, that the workings of the mine be determined, and the machinery sold off. For the information of the London shareholders, as well as to satisfy the out-adventurers in Cornish mines, that such shameful proceedings, as have occurred in Trewavas, when discovered are immediately followed by the dismissal of all who either committed or permitted the same, I may state that at the meeting alluded to, the motion of Mr Willyams was withdrawn, and in its place another submitted, to the effect that the committee and all the agents be discharged, which was carried by a large majority. This unexpected turn took place, in consequence of the discovery of the fact that the last dividend authorised by the committee was paid out of money borrowed from the bankers; and, as it appeared from the reports of several agents that under proper management the mine would in probability, become a profitable one; the ex-purser, Mr. Plomer, was reappointed, with instructions to enforce the payment of calls in arrears, and the other agencies were filled up, so that, being no longer hampered with a committee of merchants — whose only care, too often, is to sell their own goods — and, being freed from jobbing clerks, &c, better days may shortly be expected in this mine.
>
> I cannot do better than conclude, with caution to all, be aware Cornish committee mines; and to Mr. Plomer, I would say — you have now a fine opportunity improve the same.
>
> A CONSTANT READER: Truro, Feb. 3.
>
> PS. Why refer the banker's accounts for revision, any more than the merchant's?

The last paragraph is interesting, with its oblique reference to the merchants' accounts. The merchants, the people and companies that supplied the mine with everything from engine parts to cordage, were frequently shareholders in mines and not used to scrutiny of any sort on their charges.

The meeting held on March 2nd 1846 had more disappointments.[76] The accounts showed that labour and merchants' costs for November-December was £1,873 1s 1d and ores sold £759 1s 6d; total loss on two months was £1,113 19s 7d. It was then resolved to divide the loss and make a call and that losses would be divided and collected every two months. Captain R. Pope was to be appointed day and night captain at a salary of seven guineas per month; should he refuse the job was to be offered to W. Richards on the same terms. Ore bills were to be passed to the credit of the mine in the month they were received. The balance of the £3,411 4s 11d due to Messrs Ricketts and Co. was to be reduced as quickly as possible and the purser was directed to enforce, by the Vice-Warden's Court if necessary, immediate payment of arrears of calls, to be paid to Ricketts and Co.

The resolution of the previous meeting regarding the interest charged by the bank was rescinded. The purser was then directed to pay Tweedy and Co. and Willyams and Co. the money advanced by them. Mr Turner, on behalf of Ricketts and Co., had requested that the mine's account be closed. Tweedy and Co., and Willyams and Hodge were requested to take the mine's account and to advance funds for working the mine to a maximum of £2,000. However, rather surprisingly, Willyams declined to take the account on behalf of the Miners' Bank despite the previous offer.

Another meeting was held on April 21st at Pearce's Hotel, Truro.[77] The state of the mine was considered and it was resolved that the charges for interest and commission should be referred to Ricketts and Co. for revision (presumably a reduction). A further meeting was to take place on April 31st at Pearce's Hotel to adopt any measures deemed expedient. The cost now standing on the books to the end of October was £2,829 1s 6d and this was to be divided among the adventurers, to be collected immediately by the purser. Mr. Willyams gave notice that at the next meeting he would move that the mine be stopped and the materials be sold forthwith. Why Willyams should have such a sudden and abrupt volte face is not known. Be that as it may, the engine and other materials were advertised for sale the following week.

The advertisement in the *West Briton* gave somewhat more detail of the items for sale.[78] This included 150 fathoms of 'best Shropshire Iron Flat-rods,

140 fathoms of 9/16th-inch Whim Chain nearly new, several tons of Horse Whim Chains nearly new, two Horse Whims and Shaft Tackles complete' as well as the contents of the smiths' shop, this including various lengths of timber (up to 66 feet long), miners' tools and numerous miscellaneous items.

This advertisement slightly preceded the meeting held on May 4th, at which a loss of £651 11s 6d was reported.[79] This came from costs of £1,422 6s 2d versus ores raised and sold of £770 14s 8d. On commencing the meeting he stated in a circular to the adventurers:

I think it right to inform you, that 32 Shares in this Mine have been resigned, and it will be a matter for the consideration at the next Meeting whether the Mine shall continue to work in its state, or be stopped. I shall have the concern inspected by one or more Competent Mine Agents before the Meeting, in order that the Adventurers may be well advised, particularly with reference to the Western Part of the Sett, which I apprehend to be very well worthy a trial. Messrs. Tweedy & Co. have refused the Bank Account, except under the guarantee proposed by them, and to be signed by a sufficient number of Adventurers. - And Messrs. Vivian & Co. have declined it also.[80]

The mining report was presented by Captains N. and J. Vivian, indicating yet more changes.[81] The 96 fathom level, the bottom of the mine, had been driven 20 fathoms beyond the diagonal shaft; 20 fathoms beyond this a winze had been sunk below the 85 to the 96 and two fathoms driven back towards the 96 level. Another winze had been sunk beyond this and two tribute pitches were at work at the bottom. Beyond this, all levels were in poor ground 'and we do not mark any improvement in the prospects of any other operation now carrying on'. The main lode had 'most decidedly' failed at depth and the present proceeds would not match the

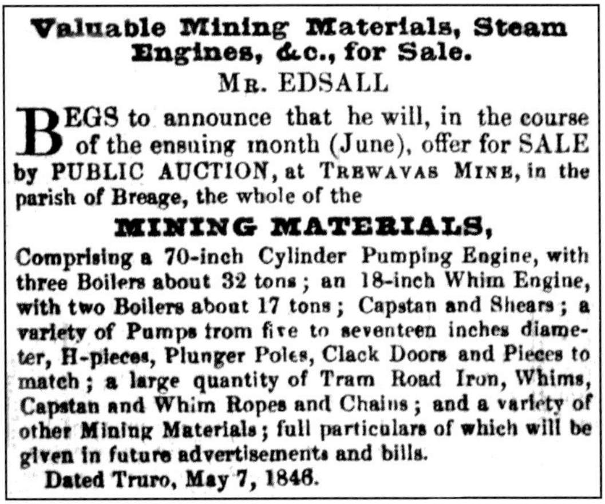

**Valuable Mining Materials, Steam Engines, &c., for Sale.**

MR. EDSALL

BEGS to announce that he will, in the course of the ensuing month (June), offer for SALE by PUBLIC AUCTION, at TREWAVAS MINE, in the parish of Breage, the whole of the

**MINING MATERIALS,**

Comprising a 70-inch Cylinder Pumping Engine, with three Boilers about 32 tons; an 18-inch Whim Engine, with two Boilers about 17 tons; Capstan and Shears; a variety of Pumps from five to seventeen inches diameter, H-pieces, Plunger Poles, Clack Doors and Pieces to match; a large quantity of Tram Road Iron, Whims, Capstan and Whim Ropes and Chains; and a variety of other Mining Materials; full particulars of which will be given in future advertisements and bills.

Dated Truro, May 7, 1846.

**Figure 26. Sale notice from the *Royal Cornwall Gazett*e of 8th May 1846.**

monthly expenditure of £250.

It was then resolved that the mine be stopped immediately and the materials sold at auction by Mr Edsell. Only the ore that could be taken away that would not delay the sale of materials would be stoped (this was usually referred to as 'picking the eyes' of the mine). Messrs Plomer, Tweedy and Kendall were appointed the committee to wind up the company and mine and all moneys received were to be paid into the Union Bank in Helston. It should be noted that there are no reports or rumours of any flooding at the mine; it closed purely for financial reasons. It was however a direct result of the mine's closure that made flooding more likely, when the directors ordered that 'only such ore as will pay' was to be removed, 'such as is left in pillars, shafts and other places, the winning of which would have endangered or obstructed normal working' (but see also below). This 'picking the eyes' of a mine was usually a last-ditch effort and would normally ensure that the mine would be less likely to work ever again. Aside from the richer ore being removed, the loss of pillars would ensure that workings would collapse; in the case of undersea workings an inundation would naturally follow which, with 19th century technology, could not be pumped dry.

Another letter regarding Wheal Trewavas (and including one of the mine's near neighbours, Wheal Penrose, which had various characteristics in common with Trewavas) appeared in the same edition of the *Mining Journal*, penned again by Henry English. This is included in full to give the reader English's feeling of frustration at some of the issues which brought down Cornish mines:

> It is at all times painful to record instances, as relates to mining enterprise, where want of application, or due diligence, on the part of those to whom the affairs of the mine, both as regards its pecuniary arrangements and its practical development and working, are confided; hence it is, that we feel regret we should be called upon to advert once again to two mining adventures in the county of Cornwall, reflecting so much discredit on the parties concerned, at the same time that the exposition is calculated materially to prejudice the interests of the mining population — who we ever bear in mind — by the distrust, we may even say the disgust, which out-adventurers, after the late expose — and we will confine ourselves to those of Wheal Penrose

and Trewavas — is so much calculated to excite. Our columns of today give the reports of the meetings of the adventurers; and, we lament to say, that while they reflect no credit on the parties managing the affairs of the mines on the 'cost book system,' some questions arise as to the course pursued, which, we are given to understand, will form subject of inquiry. Wheal Penrose, after all manoeuvres practised is 'done'. Wheal Trewavas 'ditto,' with a call of 22/ 2s per share, or thereabouts, for cost up to October — a further call of 8/ 14s 4d for November and December — and a balance against the mine, or, rather, we should say, against the adventurers, of some 1,600/ or 1,800/ at the present moment. True it is, that the machinery is estimated at 2,700/; but then again there is a balance due to the bankers, in addition to the cost, of some 1,500/, and we know not what other little out-calls may exist. We have the accounts before us; and, furthermore, have some slight notes, which we intend to review; and we hope that we shall, by our exposure of the abuses existing in the cases immediately under review, restore mining management in Cornwall to that order, or state of consistency, as meets the notions at least, as well the interests of the out-adventurers. A question arises, and to which our particular attention will be directed next week, how far the parties, who have relinquished their shares (about one-third of the mine), are further responsible? but the subject is too important to admit of a mere passing notice.

The following week a letter from 'W. R'. appeared in the *Journal*.[82] The letter mentions the false rumour of a 'bed of ore' having been found (see below), but also makes an oblique reference to the problems of having local bankers on the committee:

TREWAVAS MINE
Sir, — I have just received a letter from Helston, contradicting the report respecting the bed of ore found in Trewavas Mine. My informant acquaints me, 'the report has been examined into, and found not to be true;' so here, I suppose, ends the mine for the present adventurers. I trust, however, that we shall yet derive a something from our outlay, and that henceforth more honesty will be displayed, where management devolves on Cornishmen, and that a member of

> the committee; although he maybe a banker, will not lend himself to declaring dividends, although he may, as a banker, advance the amount necessary for such purpose.

Further in the same *Mining Journal* was another item from Henry English regarding Wheals Penrose and Trewavas. Regarding the latter he expressed his regret 'at finding so many rumours afloat, with reference not only to the mines, but reflecting on the parties on whom has devolved the management'. Concerning the rumour of ore being found at the 85 fathom level, it had been discovered that one of the tribute pitches had been 'powdered' (*i.e.* richer ore had been spread around it). It was only after the decision to close the mine had been made that the 'discovery' was made. The end of the mine was thus becoming rather chaotic.

A month later came the mine's last accident.[83] On June 4th John Russel, one of the miners, had been employed removing materials from underground. Somehow Russel slipped from where he was working, about 8 fathoms from surface, and fell 20 fathoms 'whereby his skull was so dreadfully fractured as to occasion immediate death. Verdict accordingly'.

The sale of the mining materials was advertised to take place on 23rd June.[84] This was postponed and eventually took place on July 27th, though there are no accounts of the proceedings.[85] The advert also provides the only record of an 18-inch whim, which was also for sale; possibly this was the 16-inch, after being recylindered, or just a mistake. This included 'Cylinder Case, Cage (winding drum) and Fly Wheel, and two Boilers about thirteen tons'. The 70-inch engine and some of the materials, including the account house furniture, were not sold and were re-advertised for sale on 18th August, 'with two Cast Iron Beams'.[86] A petition, presumably in the Vice-Warden's Court, of Plomer v Olney and Clouter, was dismissed in November.[87] The defendants were part-owners of a share in the mine and the prosecution was with regard to the non-payment of calls, which was not proven. The prosecution argued that the case should be dismissed with costs, the defence without costs; the Judge decided that each party would pay its own costs.

By the time the March 1847 meeting was held the materials on the mine had been sold for £2,344 10s 7d, the greater part of which had been received.[88] The winding-up committee called for the payment of all arrears of calls up to the end of February 1846 by 7th April, after which date legal proceedings were to be taken to obtain them. After this, it was hoped to settle all outstanding merchants' bills and to pay a dividend; this had not been done by November, when a London shareholder complained, 'money now being worth 9%'.

In the same month a circular was sent out to the shareholders from Robert Tweedy.[89] It is not clear what Tweedy's job was, but he may have been the auditor, certainly he was appointed to the committee to wind up the mine's affairs in May 1846. Unfortunately, little progress had been made. On November 19th the committee had met and paid a 'dividend' of 10s in the £1 to the merchants. The purser (John Plomer) had been urged to prepare a balance sheet for the shareholders; however, despite several requests by letter, nothing had materialised. Bearing this in mind, Tweedy felt obliged to tender his resignation as he could not 'consent to continue any longer connected with the winding up of a concern so long, and, as I conceive, needlessly protracted'. He then went on to suggest that a meeting of shareholders should take place as soon as possible, bearing in mind that all of the materials had been disposed of several months previous.

John Plomer's rebuttal appeared in the same *Mining Journal*. He had received applications from some of the mine's shareholders in Truro to hold a meeting, presumably at Tweedy's suggestion. Plomer stated that it was only four or five weeks since he had received all of the bills owed the company and the cost-book brought up to date, 'which Mr. Tweedy knew very well before he published his manifesto'; several sale and call accounts remained outstanding'. Plomer also stated that he had paid over all monies owed to the adventurers on November 19th and was now a creditor for arrears of salary, etc. Taking great umbrage his letter finished 'and I beg to say that, under the peculiar circumstances of the case, I think it will be incumbent on Mr. Kendall and myself to close this matter, and the outstanding accounts of certain adventurers, under the protection of a court of equity; and I beg, in conclusion, to state my regret to all parties, that we should be thus compelled to take any steps which may cause a further delay'.

A meeting of the adventurers was convened by Plomer, to be held at Pearce's Hotel, Truro, on March 23rd, to discuss the present state of the accounts and 'to consult on the best means of their being brought to a final settlement'.[90] The meeting appears to have been held at the Royal Hotel instead.[91]

The final accounts showed:

| | | |
|---|---|---|
| Labour cost for five months to July 1846 | £1440 9s 7d | |
| Merchants' bills | 727 18s 8d | |
| Ores sold (less dues) | | £2168 8s 3d |
| | £4,450 6s 4d | |
| Total debt | | £718 1s 4d |

Tweedy had told the meeting that it was his intention to resign, presumably because of his poor relationship with Plomer, however he was persuaded to remain in post, which he did. Following the meeting a circular was sent to the adventurers:

> Helston, March 26. — We beg to hand you a statement of the accounts, and copy of resolutions passed at the meeting of the 23d inst. The materials of the mine sold for £2344 10s 7d, of which the greater part has been received. We respectfully beg, in accordance with our instructions, to demand all arrears of calls and divided costs to the end of Feb. 1846, from those adventurers who have not yet paid them; and after this notice, we shall, if needful, enforce payment of the same. When this account of costs is settled, the amount of moneys received tor the sale of materials, will be duly credited to the adventurers, and we hope there will be sufficient moneys in our hands, on the final closing of the books, to make a moderate dividend; but, of course, we cannot estimate the amount of such dividend, or, indeed, proceed any further in the matter, until the accounts are closed to a given period, and all costs paid to that time. This remark has equal reference to the adventurers who retired to February, 1846. We beg to fix on Wednesday, the 7th April next, as the latest day for payment of the undermentioned sum, now due from you on account of divided costs, to the end of Feb. 1846, which we trust you will pay to your credit with the Trewavas Committee, at the bank of Messrs Vivian and Co., Union Bank, Helston, on or before the last mentioned day, as to prevent any delay, or unpleasant consequences. J. G. Plomer; R. Tweedy; John Kendall, Jun.; Committee.

So, in great chaos the mine finally closed, the subject of flooding of any sort not mentioned. The following production figures are taken from Burt *et al.*[92]

| Year | Amount (tons) | Metal (Tons) | Value |
|---|---|---|---|
| 1835 | 267.00 | 29.13 | £2,273.20 |
| 1836 | 1,339.00 | 122.84 | £11,624.43 |
| 1837 | 1,156.00 | 95.72 | £7,208.73 |
| 1838 | 1,025.00 | 87.39 | £6,675.63 |
| 1839 | 1,262.00 | 109.50 | £8,183.78 |
| 1840 | 1,849.00 | 147.72 | £11,609.25 |
| 1841 | 2,366.00 | 176.18 | £15,659.93 |
| 1842 | 2,775.00 | 209.87 | £16,288.43 |
| 1843 | 2,316.00 | 198.35 | £15,023.18 |
| 1844 | 1,607.00 | 127.27 | £9,283.50 |
| 1845 | 977.00 | 73.15 | £5,452.50 |
| 1846 | 397.00 | 33.38 | £2,330.73 |
| Total | 17,336.00 | 1,410.50 | £111,613.29 |

Canon Rogers, the ever-patient owner of the land mined by Wheal Trewavas, was at this time (April 1846) having problems with Wheal Penrose, the eastern near-neighbour of Trewavas. It appears that Rogers had advanced the capital for all of the materials and equipment at the most recent re-opening of that mine. With no obvious sign of a repayment Rogers had by now resorted to solicitors in an attempt to retrieve the money he had advanced. The mine's machinery and materials were advertised for sale in May 1846, partly because the lode mined had failed, and partly in order to pay Rogers back.[93] Interestingly the purser of Wheal Penrose was one John Plomer, lately purser of Wheal Trewavas. Another two familiar names were Messrs. Turner and Willyams, both bankers and committee members.

The death of Captain Thomas Teague was announced in June 1857.[94] While he had been associated most with Tresavean Mine near Redruth, he had been associated with Wheal Trewavas in its early days.

Captain William Martin, another of the mine's agents, died on 3rd July 1861, aged 72.[95] Born in 1789 he commenced his mining career as a barrow-boy and then as a tributer. He first became a mine captain at the young age of 23, while four years later he applied for, and obtained, the sett of Tresavean Mine where he was appointed the lord's agent. Opening up Tresavean led to the opening of Trethellen, Brewer, Barrier and Treviskey mines and he was also credited with the discovery of Wheal Trannack as well as Wheal Trewavas. His funeral on 7th July was attended by over 2,000 people.

**References**


1. English, H., 1826. *A Compendium of Useful Information Relating to the Companies Formed for Working British Mines* [etc.], Boosey & Sons, London, pp15-18, 95; Smallwood, G., (ed.) 1825. *The Monthly Repository of Theology and General Literature*, January to December, Inclusive, volume XX, London p502
2. Watson, J. Y., 1843. *A Compendium of British Mining*, London, p46 (confusingly this is in the section titled *The Mines of St Just and St Ives*)
3. Cornwall Record Office: document STA 711/20
4. *Mining Journal* 17.10.1835 p61
5. *Royal Cornwall Gazette* 4.7.1835 p2 (the first sale was 43 tons)
6. *Mining Journal* 12.9.1835 p24
7. *Mining Journal* 21.11.1835 p111; *Mining Journal* 16.1.1836 p27
8. D. B. Barton, 1989. *The Cornish Beam Engine*, Cornwall Books, p173
9. *Royal Cornwall Gazette* 19.2.1836 p3
10. *West Briton* 8.4.1836 p3
11. *Royal Cornwall Gazette* 27.5.1836 p3
12. *West Briton* 3.6.1836 p3
13. *Royal Cornwall Gazette* 23.9.1836 p3
14. Lemon, C., 1838, The Statistics of the copper mines of Cornwall, *Journal of the Statistical Society of London*, p78
15. *Royal Cornwall Gazette* 16.9.1836 p3
16. *Public Ledger and Daily Advertiser* 13.10.1865 p4
17. *Royal Cornwall Gazette* 11.11.1836 p3
18. *Royal Cornwall Gazette* 11.11.1836 p3
19. *Royal Cornwall Gazette* 2.12.1836 p3; *Bristol Mercury* 26.11.1836 p3
20. *Mining Journal* 13.5.1837 p165
21. *Mining Journal* 12.5.1838 p147

22. Royal Institution of Cornwall document HJ/1/18
23. For more information on Pascoe Grenfell see the articles by Penny Watts-Russell in the *Journal of the Trevithick Society*, 2015 pp15-43, 2016 pp36-62 and 2017 pp29-63
24. *Mining Journal* 1842 p148
25. Henwood, W. J., 1843. On the Metalliferous Deposits of Cornwall and Devon; *Transactions of the Royal Geological Society of Cornwall* V, pp1-387; Vibert, Penzance
26. Cornwall Record Office document X154
27. *Royal Cornwall Gazette* 3.4.1840 p3
28. *Royal Cornwall Gazette* 10.4.1840 p3
29. *Royal Cornwall Gazette* 24.4.1840 p3
30. *Mining Journal* 2.5.1840 p142
31. Cornwall Record Office document J1308/28-30,35,37
32. Henwood, W. J., 1843. Observations on Subterranean Temperature in the Mines of Cornwall and Devon (Appendix No. 1). *Transactions of the Royal Geological Society of Cornwall*, volume 5 p392: Vibert, Penzance and Longman, Brown, Green & Longman, London
33. *Mining Journal* 24.10.1840 p342
34. *Royal Cornwall Gazette* 29.1.1841 p3
35. Royal Institution of Cornwall document: HJ/1/19
36. *In:* Pole, W., 1844. A Treatise on the Cornish Pumping Engine; in two parts. John Weale, London, pp141, 144
37. *In: Mechanics Magazine, Museum, Register, Journal and Gazette*, January 1st – June 25th 1842, vol. XXXVI. p192 London
38. *Royal Cornwall Gazette* 11.3.1842 p3
39. *Royal Cornwall Gazette* 2.12.1842 p2
40. *Penzance Gazette* 14.12.1842 p4
41. *Royal Cornwall Gazette* 3.4.1843 p3
42. *Royal Cornwall Gazette* 20.10.1843 p3
43. Watson, J. Y., *op. cit.* p46
44. *Mining Journal* 17.2.1844 p57
45. *Mining Journal* 11.5.1844 p158
46. *Royal Cornwall Gazette* 14.6.1844 p3
47. *Mining Journal* 29.6.1844 p220
48. *Mining Journal* 6.7.1844 pp228, 231
49. *Mining Journal* 13.7.1844 p237

50. *Mining Journal* 7.5.1836 p161
51. *West Briton* 26.7.1844 p3
52. *Royal Cornwall Gazette* 13.9.1844 p3
53. *Royal Cornwall Gazette* 11.10.1844 p3
54. *Royal Cornwall Gazette* 29.11.1844 p3
55. *Mining Journal* 10.8.1844 p270
56. *Mining Journal* 14.10.1844 p265
57. *Penzance Gazette* 30.10.1844 p2
58. *Mining Journal* 2.11.1844 p379
59. *Mining Journal* 4.5.1845 p121
60. *Mining Journal* 14.6.1845 p301
61. *Royal Cornwall Gazette* 5.9.1845 p3
62. *Mining Journal* 18.10.1845 p581
63. *Mining Journal* 8.11.1845 p633
64. *Penzance Gazette* 12.11.1845 supplement p1
65. *Mining Journal* 15.11.1845 p648
66. *Mining Journal* 12.1.1846 p14
67. *Mining Journal passim*
68. Cornwall Record Office document: CMDA/10/3
69. *Mining Journal* 19.1.1846 p29
70. Barclays Bank Archive https://www.archive.barclays.com/items/show/5224
71. *Mining Journal* 19.1.1846 p29
72. *Mining Journal* 19.1.1846 p29
73. *Mining Journal* 24.1.1846 p30
74. *Mining Journal* 7.2.1846 p59
75. *Royal Cornwall Gazette* 6.2.1846 p3
76. *Mining Journal* 8.3.1846 p113
77. *Mining Journal* 31.4.1846 p47
78. *West Briton* 8.5.1846 p3
79. *Mining Journal* 9.5.1846 p197
80. Cornwall Record Office Brook Index
81. *Mining Journal* 9.5.1846 p197
82. *Mining Journal* 16.5.1846 p208
83. *Royal Cornwall Gazette* 12.6.1846 p3
84. *Royal Cornwall Gazette* 19.6.1846 p1
85. *Royal Cornwall Gazette* 8.5.1846 p3

86. *Royal Cornwall Gazette* 14.8.1846 p3
87. *Royal Cornwall Gazette* 20.11.1846 p4
88. *Mining Journal* 3.4.1847 p149
89. *Mining Journal* 13.3.1847 p114
90. *Mining Journal* 20.3.1847 p123
91. *Mining Journal* 3.4.1847 p149
92. Burt, R., with Burnley, R., Gill, M. and Neill, A., 2014. *Mining in Cornwall and Devon: Mines and Men*. University of Exeter Press, Exeter.
93. *Mining Journal* 9.5.1846 p197
94. *Cornish Telegraph* 20.6.1857 p1
95. *The Gentleman's Magazine and Historical Review* by Sylvanus Urban, Gent. John Henry and James Parker, London. Volume XL, July-December 1861 pp333, 334

# Chapter 2

## Later Attempts

While Wheal Trewavas closed in 1847 it was to reappear a number of times after that date in a series of unsuccessful trials. In 1851 an advert stating prospectuses for the Trewavas Copper and Tin Mine could be obtained from Messrs. Wilkinson, Gurney, and Stevens of Lombard Street, London, appeared.[1] In March 1857 18 tons of copper ore were sold for £68 18s, probably from the dumps. In July 1858 an item in the Wendron Consols' accounts for 'Mrs Agar for Sett of Trewavas' for £70 almost certainly refers to the site near Helston.[2] In February 1863 it was rumoured that the mine was about to be reopened, and that £750 worth of copper ore had been sold in the last month of the working.[3]

In June 1864 James Lane, a London sharebroker, noted that it was about sixteen years since the sea 'so mysteriously broke in'[4] while in 1867 the Trewavas Tin and Copper Mine was floated.[5] Under Captain J. R. Riddington this operation was in 256 shares of 5s each; dues from Canon Rogers were 1-18th. The object was to extend the adit on the Nimble Cutter Lode to cut the Great Tin Lode running east-west through the sett.

The more grandiose sounding Trewavas, Tin, Copper and Silver-Lead Mine commenced operations in April 1874.[6] The two principals were S. S. Cowper and Captain C. S. St George; the manager was Captain William Treweeks. Once again the bankers were Messrs Tweedy, Williams and Co. The capital of the company was 6,000 shares of which 3,000 were being offered. The sett was said to measure 1,200 yards north-south and 800 yards east-west. Of the shares, Cowper and St George held 1,000 each. A favourable report had been given by Captain Michael Williams, who had been an agent for eleven years in the old mine. According to Williams, the workings ended 'consequent upon working under the sea, where, from want of care in opening the upper levels, works were conducted too near to the sea bed, which, with defective ventilation, led to a stoppage of the mine'. The mine had not been reworked 'from want of capital and various other causes'.

Because of this it was decided to ignore the submarine section and concentrate on the 'equally rich western [copper] lodes in conjunction with the tin'. Favourable reports were given by captains Michael William Martyn, the former agent, Charles Thomas of Dolcoath, Thomas Richards late of Prosper United and Edward Blewitt (mine not given). Richards recommended extending the adit in that direction on the Nimble Cutter Lode. Two of the tin lodes were reported as having been extensively stoped down to the water level; £1,000 was estimated as required to open the mine and £4,000 to bring it to a dividend paying state. The start-up costs were inclusive of an engine of 20 to 24-inch cylinder. The shares were reportedly over-subscribed four days before the list of applicants was due to close. In June 1874 it was stated that proper surveys and plans were being prepared, however little progress was actually recorded in the *Mining Journal*.[7]

In October 1875 'Shepherd's Bush' complained about companies brought out with insufficient capital, including '... Trewavas, where a Mr Troop, of Helston, who was secretary, appears to have dropped out of sight altogether'.[8] Kelly's 1873 *Post Office Directory* shows that Alex. Troop, 'gardener, etc' lived in Coinage Hall Street. Unsurprisingly this operation also came to nothing,

In August 1876 it was reported that yet another company had been formed to work the sett.[9] This was under the management of Captain John Curtis, mineral agent to the Duke of Leeds and the former agent of Leeds and St Aubyn Mine.[10] Nearby the Mounts Bay company was said to be making

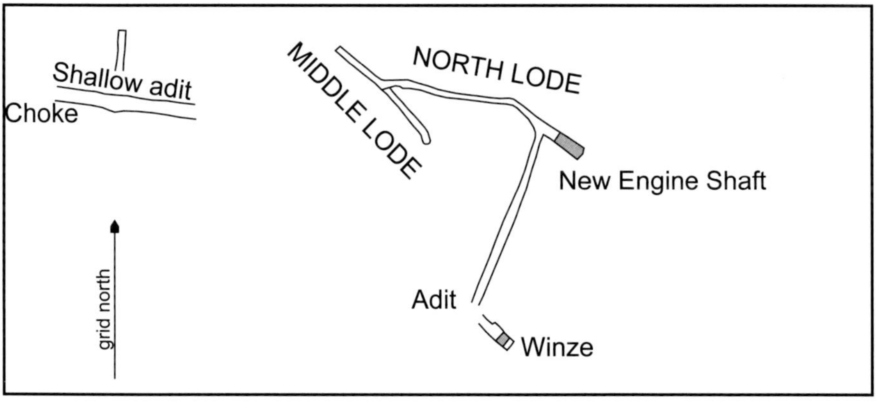


**Figure 27. Plan of the adits at New Engine Shaft surveyed by Ali Neill and Kevin Baker. Scale not known.**

good progress on the 'Trewavas lode'. In November Curtis stated that he was sanguine that a good mine would be found as some beautiful stones of superior quality yellow copper ore (chalcopyrite) had come from the end of the adit.[11]

During July 1876 the Curtis reported that a ladderway was being put in the Engine Shaft down to adit level, which was cleared and driving commenced west on the Old Trewavas Lode. In August it was stated that the old burrows had been set to dress for copper at 16s in £1 tribute. A puffing article in September stated that dead ground being driven through was a valuable china stone worth £60,000 or upwards. Captain Curtis's reports between July and October suggest the only real work was continuing to drive the adit, the lode at first containing some mundic (pyrite, possibly with some arsenopyrite) but later was stated to contain copper ore 'from top to bottom of the end', an assay by a Professor White giving over 20% copper. A small parcel of copper ore was sold in September probably by private contract. In October it was stated a Mr Thompson of Crown Court, Threadneedle Street, London held a large interest in the company and was prepared to deal in the shares, but it seems that soon afterwards operations ceased. Thompson was to feature prominently in the proceedings of New Penrose company, four years later. Trewavas continued to be included in the *Mining Journal* share lists until August 1877.

In January 1878 an advert for an 'authorized issue of 600 £10 debenture 1st mortgage bonds' was issued for half the capital of Trewavas (Mounts Bay) Mine.[12] This included all of the machinery and plant although none was described; there was likely to have been very little there in any case, given the small-scale nature of the preceding operations. The debentures guaranteed a 10% return on profits. An extravagantly optimistic letter from Captain Joseph Vivian, 'late of Great Wheal Fortune and many other mines' (including, briefly, Trewavas) stated that:

> These crosscuts will undoubtedly open up some very rich courses of ore, and if driven, will cut the lodes at twelve different points; should only one half of these points turn out to be productive, your property would easily sell for £50,000. *There is no doubt on my mind that all of the points will be found productive.*

In April the company was calling itself The Mounts Bay Company (Limited) with a capital of £12,000 in £1 shares; 6% interest would accrue to the shares.[13] There was also to be an issue of 100 £25 7% first mortgage bonds, redeemable on 3rd April 1883 and interest payable on 3rd October and 3rd May annually; the first payment was to be made on 3rd May 1883. In addition to Trewavas it was to work Prospidnick Granite Quarry, the Gwealhellis and Gwealfolds sett at Helston and 'others elsewhere'. The other reports attached were essentially those from the Trewavas, Tin, Copper and Silver-Lead Mine prospectus with the addition of the one from Captain and one from John Curtis.

The purchase price was to include the cost of driving the deep adit 70 fathoms west from the cliff. Carrying this out would reach ore of sufficient value that 'immediate profits' of £6,000 to £10,000 per annum would be made, and this would increase as more levels were opened. The report from Curtis indicated that work had already commenced: the engine shaft was secured to the deep adit and had been driven 40 fathoms west. The value of the lode had been slight but had been increasing; it was presumed that a valuable deposit would be found when the lode intersected a flookan about 30 fathoms beyond the present end. Once again the reason for the mine's earlier closure was stated that the workings were too close to the sea-bed.

In March 1879 another attempt to resume operations was made, under the auspices of Captain John Curtis.[14] A few shareholders had acquired the property and a search had commenced to locate and develop the lodes westwards and inland. Locally it seems that nothing was known of the operations. In December the *West Briton* stated:

> Trewavas is a copper mine, which ceased to work about twenty-five or thirty years ago. It yielded some profit, I understand, and worked for several years. As the sea has found ingress into the workings, resumption of operations must be difficult, if ever attempted. A piece of ground at the west, in the same sett, is considered a fit site for operations.[15]

In 1880 an attempt was made to rework the mine under the name of the New Penrose Tin and Copper Mining Company Limited.[16] This company had a capital of £12,000 in £1 shares, of which two-thirds had already been applied for. The directors were Major the Hon. O. G. Lambert, John Long and H. Stephenson of London and William Thomson of Plymouth, who was the managing director. Captains Curtis and Vivian were said to have reported favourably on the sett. No results of this attempt are known, though they were surely unsuccessful, as there are very few references to this operation and no recorded production figures. An enticement to potential shareholders Vivian was quoted as saying, with great emphasis on values, 'It is not rare in this district for courses of ore to be met with worth 500/ (five hundred pounds) per fathom. I have seen courses of ore in Great Wheal Vor worth 800/ (eight hundred pounds) per fathom'. Work on the mine had commenced in June.[17]

The first statutory meeting was held at the Angel Hotel in Helston on 5th February 1881, chaired by William Thompson, at which a report from John Curtis was read.[18] Since the previous June only 9 fathoms and 2 feet had been driven on the various lodes and cross courses while a winze had been cleared and secured for 16 fathoms below the shallow adit to communicate to the deep adit. A new footway with ladder sollars had also been fitted. The deep adit, on the 'old Trewavas copper lode', had been extended 7 feet west of the winze on a lode 15 inches wide. In the winze the lode was 18 inches wide, carrying yellow and black copper ores.

All the stuff from driving was dumped into the Old Engine Shaft to save hoisting to surface. By driving about 30 fathoms further west this level was expected to intersect a flookan course, a little further west of which former adventurers had sunk a shaft 18 fathoms; this was 'to have been an engine-shaft or the most suitable place for an engine'. Thompson stated that driving east and west on this flookan course would intersect the other lodes in the property, seven in number. These were all statements which had, more or less, been made during the previous company's working. The mine was now laid open and well-ventilated and Curtis recommended putting more men in to drive east and west; no money was required for horse hire for hoisting as all of the stuff (presumably meaning waste) was being put into the Old Engine Shaft.

At this time a private prospectus was prepared, in order to issue unallotted shares at par.[19] Thompson was hoping to form a board consisting of himself, Mr. Criper of Tavistock, and he hoped Mr St. Aubyn MP and a representative of Harvey & Co. of Hayle. The prospectus included amongst various reports, those which had first appeared in 1874. The objects included putting up an engine on the shaft west of the flookan course to explore the various lodes below adit.

The following month a further prospectus was published,[20] however the list of directors was completely different, comprising William Trevenen (banker and Mayor of Helston), Thomas E. Owen of Plymouth and late mayor of Totnes, William Criper (who later that year became a director of the East Herodsfoot Silver-lead Mine[21]), George Eustice of Hayle, and Thompson. The mine was 'pushed on with all vigour' and the deep adit was now 80 fathoms from the cliff.

A rather telling paragraph appeared in the *Cornishman* in July:

> The New Penrose Tin and Copper Mining Company have commenced operations on the Trewavas Head, but the Company lacks spirit in connection with the reworking. Messrs. Wm. Thompson and Co. .... of Plymouth, are the proprietors, and it is much to be desired that they will give the sett a fair trial. Experienced miners speak highly of the lodes. A house has been built and exploration on a limited scale conducted.[22]

Driving of the adit continued throughout the year, reports at times stating the lode contained some copper ore and mundic, however many of the reports comprise one or two lines only.[23] In August work started to secure the collar and clear the shaft to the west which was named Engine Shaft. This had been flooded to about 5 fathoms below surface, but it appeared the water in this was gradually being cut down by the adit progressing towards it. A new tin lode was reported to have been discovered in that month, likely 'to materially enhance the value of the shares'. The shaft was cleared to 12 fathoms from surface in October[24] (although the same newspaper also states that operations were being carried out on a limited scale by four workmen[25]) and to the bottom in November, and Curtis's report dated 29th November referred to driving a 17 fm level east and west of the shaft. The lode in the shaft was stated to contain a little tin as well as some copper ore, and tin was also reported in the lode in the adit at this time. The last report seen was dated 12th December 1881.

On 9th March 1882 a Chancery Case was heard, Molony v. Thompson.[26]

Major General Charles Preston Molony was the registered holder of 400 shares; he wished to have his name cancelled as a shareholder and the £400 he paid for them to be returned. The grounds for this was that he had been induced to take them by fraudulent statements, 'it being alleged there was in fact no mine, and that the company was in reality a creation of a Mr. Thompson, a stock and share broker, of Plymouth'. William Thompson was described as 'the managing director, secretary, and treasurer, and, in fact, the company itself'. Molony, 'an Irish gentleman', was no doubt trying to escape any further liability for the uncalled capital on his shares, particularly in the event of the company failing, which must have appeared imminent. He alleged it was untrue that half the capital had been taken up or that a board had been formed, and also stated that some of the meetings were only attended by Thompson. It was further claimed that of the directors listed in the prospectus, all except Trevenen had repudiated any connection with the company. On the side of the company all these claims were denied, but it transpired Molony's shares were issued before the company was actually registered, so an order was made to remove his name and refund the money he had paid with interest.

A meeting was held on 31st March on Molony's petition, before Sir Charles Hall, at which it was decided to wind up the company.[27] Molony, of the Indian Army, was one of the mine's creditors. John Henry Tilly was appointed Official Liquidator on 8th May, when the case was heard once again before Sir Charles Hall, at the Royal Courts of Justice.[28]

The Inspectors of Mines Reports give employment figures:

| | |
|---|---|
| 1880 | 2 males aged above 16 underground. |
| 1881 | 4 males aged above 16 underground. |

The final mention of possible operations at the mine is a report in the *West Briton* in 1907 stating that J. P. Rogers was said to have made an application for the lease of the mine (spelled Treworvas).[29]

In 1961 an article by 'the well-known local authority, Mr P. Cowls,' without dates or references, appeared in *Old Cornwall* magazine regarding the flooding of the mine.[30] According to the author:

It happened the day that the 'tributers' annual dinner was to take place

> with the meal being served underground 'well out under the sea'. Two men had gone underground to put the final touches to the tables of food ready for the dinner in the evening. One of them noticed some water leaking from the roof of the tunnel and dripping onto one of the tables. The two men made a hasty exit to the surface and soon, before the time planned for the dinner, the sea had broken in leaving the dinner for the fishes. As the mine was irrevocably breached below low water all seaward work at this mine ceased henceforth.

Certainly this information is not borne out in any of the information so far accumulated regarding this mine, and has been questioned by many other authors; the 1874 report by Captain Michael Williams makes no reference to any such incident. The actual circumstances are somewhat odd, particularly with regard to a 'tributers' dinner' (one wonders if this is a misinterpretation of a shareholders' dinner, and holding such an event underground, particularly at some distance beyond the cliff line, sounds very strange indeed. The space necessary to hold such a dinner is unlikely to have been available, even in one of the stopes. It is also assumed that such an inundation would surely have been reported in the local newspapers, which it was not.

Regardless of how and when, the mine is now flooded. This would have happened without a marine incursion simply from the fact that any hole in the ground below the water table will fill up. The fact that towards the end of the mine's life ore was being extracted as quickly as possible suggests that the miners may have worked with less caution than absolutely necessary, resulting in a situation similar to that of the '40 backs' at Levant, where miners had followed ore so far above the back of the level that the mine eventually flooded.[31] Fortunately this occurred several years after the mine closed, although it appears to have happened abruptly.

In August 1959 the late Justin Brooke spoke to Harry Basher of Rinsey;[32] the latter told the story that the sea broke into the mine after it had been closed, a rise having been extended by tributers too close to the sea bed. Canon Rogers was said to have been enjoying the dues on copper ore raised from under the sea, which should have been paid to the Duchy (the Duchy of Cornwall was entitled to all dues on minerals raised from beyond the low-water mark; the Crown was entitled to the dues from minerals raised between low and high-water), but before the Duchy agents could inspect the mine, it was closed down and the water was allowed to rise.

In support of this story the Duchy of Cornwall was beginning to claim dues from submarine mines from 1841.[33] This greatly affected the St Just area and was commented on the following year.[34] However the *Royal Cornwall Gazette* in 1859 also mentioned that the mine had 'been abandoned in consequence of a dispute between the Lord and the Duchy', which adds a little more credence to the story.[35] It does seem a somewhat vindictive reason to close a mine, particularly for a man of the church, and his previous generosity regarding the mines he was associated with makes this seem unlikely.

Miscellaneous production[36]

| Year | Amount (tons) | Metal (Tons) | Value |
|---|---|---|---|
| 1857 | 18.00 | 0.72 | 68.40 |
| Total | | | |

**References**


1. *Mining Journal* 19.4.1851 p181
2. *Royal Cornwall Gazette* 2.7.1858 p7
3. *Mining Journal* 14.2.1863 p116
4. Alasdair Neill, personal communication
5. *Mining Journal* 30.3.1867 p208; *Cornish Telegraph* 3.4.1867 p3
6. *Royal Cornwall Gazette* 2.5.1874 p8
7. *Mining Journal* 11.4, 25.4, 9.5 and 6.6.1874
8. Alasdair Neill, personal communication
9. *Royal Cornwall Gazette* 19.8.1876 p5
10. *West Briton* 19.11.1874 p3
11. *Royal Cornwall Gazette* 11.11.1876 p5
12. *Dublin Evening Mail* 1.1.1878 p1
13. *Cornish Telegraph* 2.4.1878 p2
14. Alasdair Neill, personal communication
15. *West Briton* 18.12.1879 p6

16. *The Standard* 7.7.1880 p7
17. *Cornishman* 15.7.1880 p8
18. *Western Daily Mercury* 9.2.1881 p4
19. Cornwall Record Office: H/157/230-231
20. *Western Daily Mercury* 28.3.1881 p1
21. *Western Daily Mercury* 14.5.1881 p3
22. *Cornishman* 21.7.1881 p5
23. *Mining Journal* 1881 *passim.*
24. *Cornishman* 20.10.1881 p5
25. *Western Morning News* 15.3.1882 p4
26. *London Gazette* 7.4.1882 p1603
27. *Western Morning News* 1.5.1882 p1
28. *Royal Cornwall Gazette* 26.5.1882 p5
29. *West Briton* 31.1.1907 p6
30. *In*: Jenkin, A. K. H., 1962. *Mines and Miners of Cornwall*, iv: Penzance – Mounts Bay, Truro Bookshop, pp45-46
31. Corin, J., 2013. *Levant, A Champion Cornish Mine*; edited and expanded by Peter Joseph, The Trevithick Society, pp56-60
32. Cornwall Record Office document: The Brooke Index
33. Penzance Gazette 28.4.1841 p4
34. Buller, Rev. J., 1842. *A Statistical Account of the Parish of St Just in Penwith in the County of Cornwall* [etc], Rodda, Penzance, p16
35. *Royal Cornwall Gazette* 9.12.1859 p8
36. Burt, R., with Burnley, R., Gill, M. and Neill, A., 2014. *Mining in Cornwall and Devon: Mines and Men*. University of Exeter Press, Exeter

# Chapter 3

# Archaeology

---

The following two chapters rely heavily on reports of the Cornwall Archaeological Unit, with some slight modifications by the author; the work of CAU in recording this enigmatic site is gratefully acknowledged.

> At Trewavas Head the roofless and windowless houses and the crumbling stacks are perched on the very brow of the cliff...
> *Cornish Telegraph* 24.7.1877 p2

Wheal Trewavas lies predominantly on a south-facing area of coastal slope. The cliffs are now given over to wildlife and tourists, and the various noises from the mine – from men and women hammering, ropes and flat-rods creaking, steam hissing and the rumble of kibbles in the shafts – have gone, replaced by the softer chatter of the visitors and the cries of the gulls and choughs. As with so many Cornish mine sites, comparatively little now remains; the present features comprise two engine houses, two stacks, two capstan plats, a number of horse whim plats and the remains of an inclined tramway. Missing are two other engine houses, the count house, one other building, the spalling floors where the copper ore was broken, various reservoirs and, of course, all of the timber structures that would have dotted the site (see figure 28). Several shafts are known as well as some possible lode-back workings.

All surviving built features at Wheal Trewavas are listed Grade II, while the site itself is a Scheduled Ancient Monument. Remarkably, for reasons currently unknown, English Heritage does not sub-divide time periods after the Medieval period and so all of the features after this are recorded as being 'Post-Medieval'. Hence, at Wheal Trewavas, and at all other similar sites in the UK, the engine houses are dated as Post-Medieval, despite the fact that no pre-Medieval engine houses occur in the UK, or, indeed, the world. It is hoped that this situation will be resolved eventually.

To the west of the mine site can be seen some more modern archaeology, the remains of a World War II Chain Home Low radar station. This opened on 12th April 1941 and closed in May 1945 and was used to detect aircraft flying under 1,500 feet as well as shipping. A little farther west was an associated mast. The radar operated on 1.5m wavelength and was manned by the RAF personnel who were accommodated in huts near Trewavas Farm. Only the station building's concrete bases survive today

The eastern pumping engine house is one of the oldest in Cornwall. This engine house is not well built, possibly because of its early date; there is no cylinder door and the seaward windows have been blocked.

A report in the *Mining Journal* states that this engine house was in use in by October 1834. Its 45-inch pumping engine had been built in Hayle by Harvey & Co. and was pumping from the adjacent New Engine Shaft. Attempts were made to sell the engine several times in the last quarter of 1844 following the acquisition of a 70-inch engine; it was probably sold in December. Unusually the cylinder bedstone remains in the house. This was usually sold with the engine and only left in place if the cylinder was scrapped; the reason it remains is unclear but, given the position of the house, it may just have been too difficult to remove. A capstan is marked on the old Ordnance Survey 25-inch map some 50m WNW of the engine house; this survives adjacent to the path leading to the engine house.

The boiler house lay to the south-east and was offset from the engine house to follow the contour of the land. It is 9.5m long by 2.4m wide and has a detached granite chimney to which it was formerly linked by a flue. Both this (now demolished) boiler house and a further building platform beyond have retaining walls up to 5m high on the seaward side. West of the shaft are horse whim and capstan platforms. By 1841 power was taken from the engine by flat-rods, most likely used to pump from Old Engine Shaft. These were probably accommodated in a shallow tunnel through the clifftop, west of New Engine Shaft.

The western engine house was built for a 70-inch engine which was in full operation by August 1844, acquired to replace the 45-inch engine. The new engine pumped from the adjacent New Engine Shaft but it is not known when the engine appeared on the mine. By October the following year the mine had paid over £2,000 for the engine and other materials.

The SSW wall of the engine house bears traces of the roof of the boiler

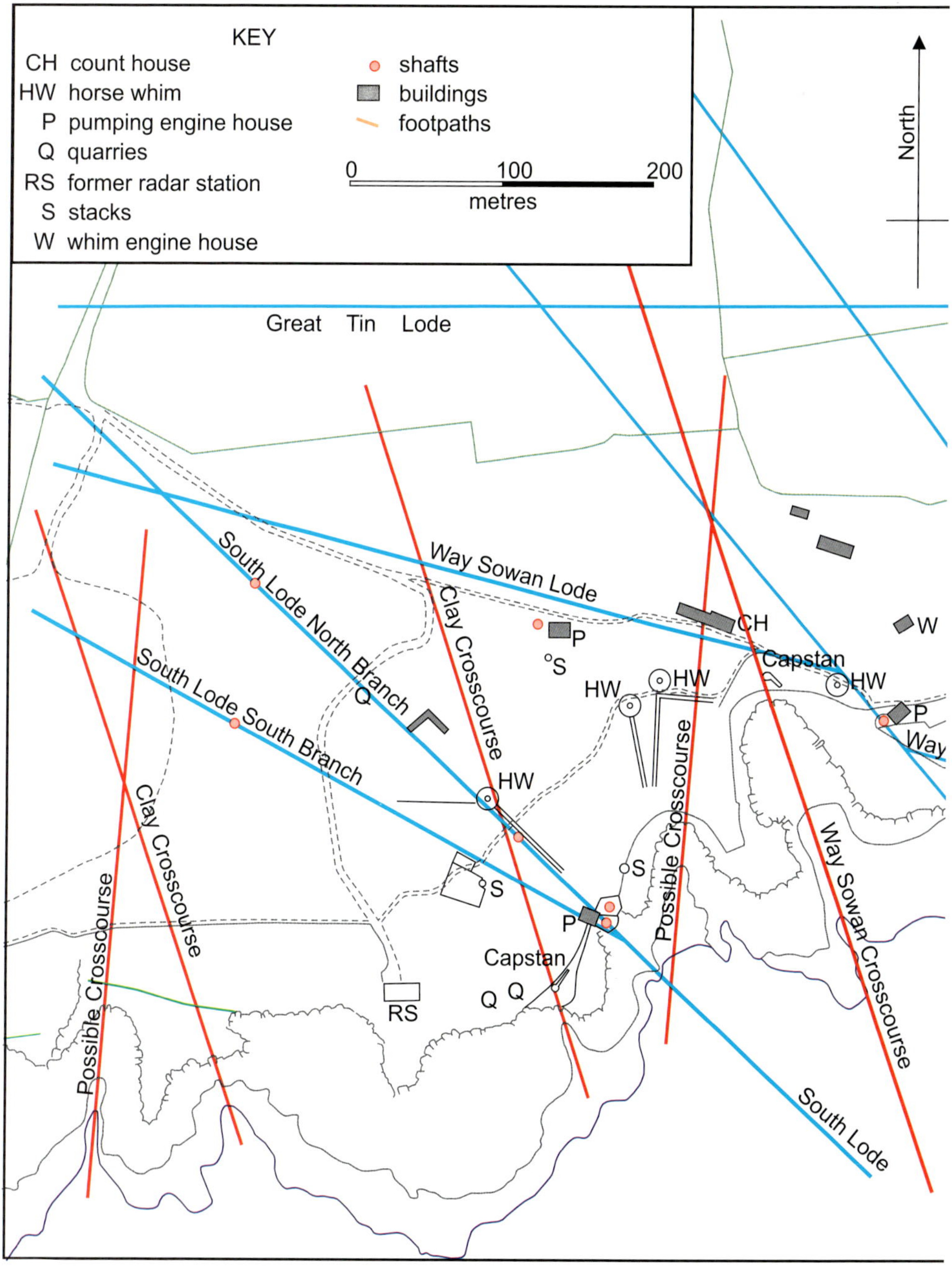
KEY
CH count house
HW horse whim
P pumping engine house
Q quarries
RS former radar station
S stacks
W whim engine house
shafts
buildings
footpaths
0
100
200
metres
North
Great Tin Lode
Way Sowan Lode
South Lode North Branch
South Lode South Branch
Clay Crosscourse
Possible Crosscourse
Way Sowan Crosscourse
South Lode
Capstan
CH
HW
P
S
Q
RS
W
Way

Figure 28. Site plan, redrawn with some alterations by the author, from the Cornwall Archaeological Unit's plan.

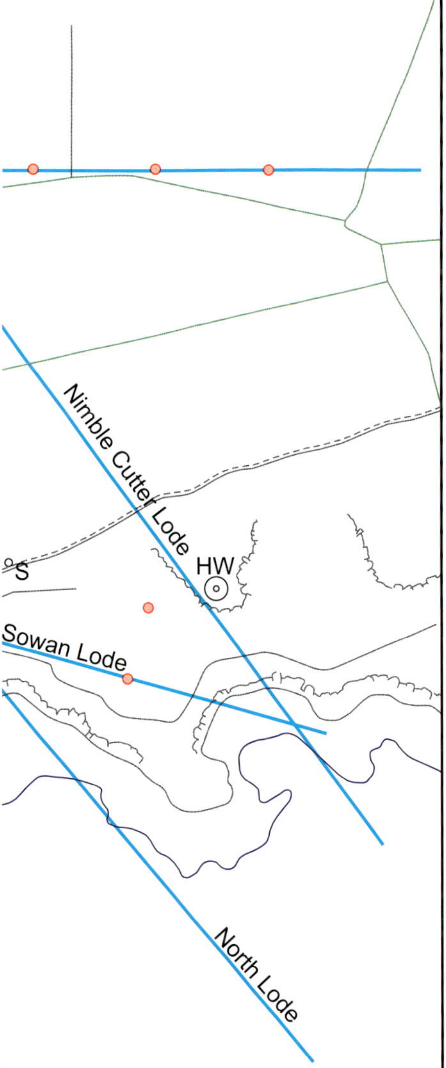


house, also visible as a platform with rubble walling surviving up to 3m high. This house extends 8.5m NNE-SSW and an estimated 16m WNW-ESE. Its rubble stone chimney is sited on the slope above. The connecting flue has rubble walling, and was roofed with stone slabs, some remaining in situ. Above the chimney is a rock-cut, perhaps for coal storage, with evidence for a structure on its northern side. The yard contains the concrete footprints of structures associated with the WWII radar station. The station itself, situated on a prominent spot to the south west, survives as a concrete base, with traces of walls, openings and wooden flooring.

The engine was offered for sale, upon the closure of the mine, in May 1846. It has been suggested that this house was built for the 16-inch whim but it was clearly built for a pumping engine, with a bob wall adjacent to the shaft.

Just west of the pumping engine house can be seen a large and exceptionally well-preserved capstan platform. A capstan was a windlass, used for manually hoisting materials such as cast-iron pumps and timber pump-rods, as well as bulk materials such as timber used underground, up and down the shaft. This is without a doubt the best preserved in Cornwall.

The platform has been used by RNAS Culdrose for helicopter landing practice. The helicopter with the World Heritage Site international assessor on board

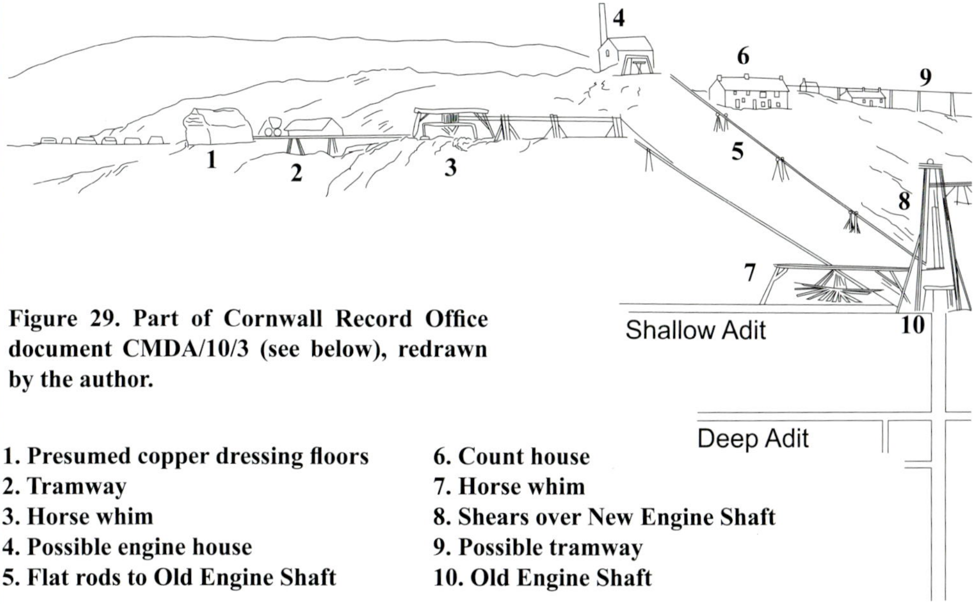


**Figure 29. Part of Cornwall Record Office document CMDA/10/3 (see below), redrawn by the author.**

1. Presumed copper dressing floors
2. Tramway
3. Horse whim
4. Possible engine house
5. Flat rods to Old Engine Shaft
6. Count house
7. Horse whim
8. Shears over New Engine Shaft
9. Possible tramway
10. Old Engine Shaft

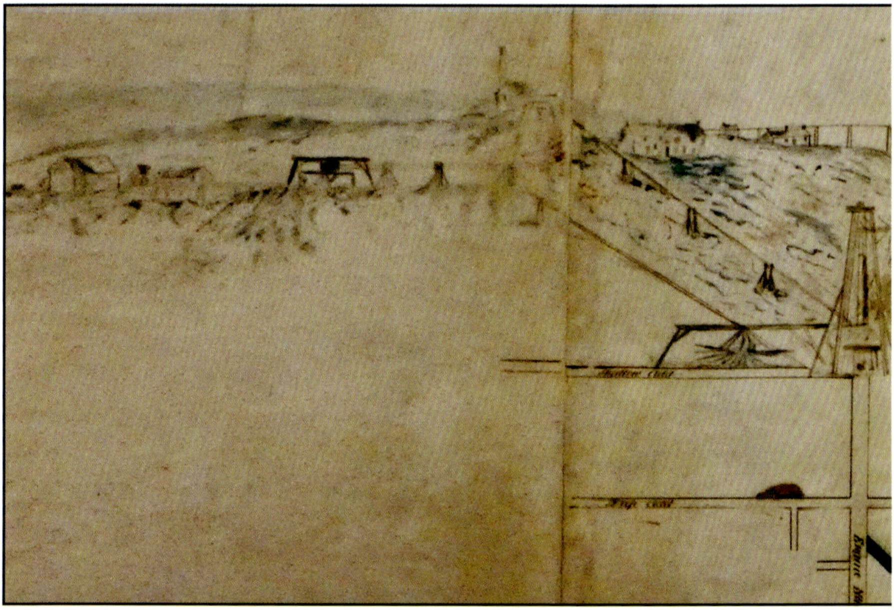

**Figure 30. The original part of Cornwall Record Office document CMDA/10/3.**

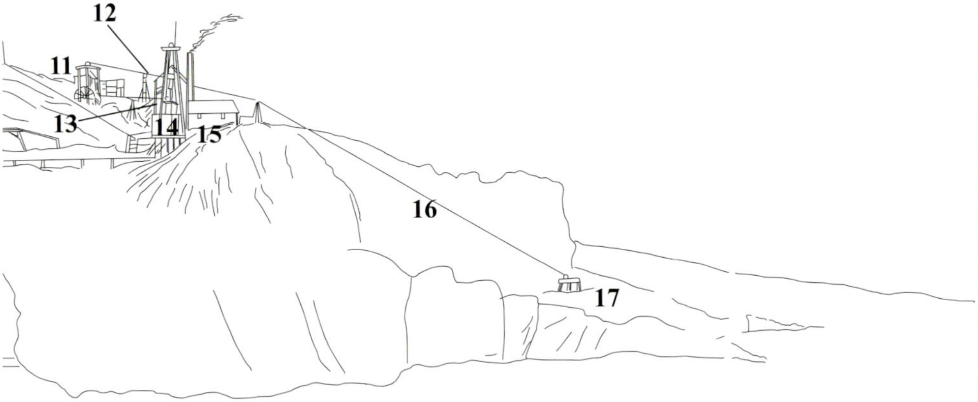


11. Presumed steam whim
12. Whim rope
13. New Engine Shaft engine house
14. New Engine Shaft
15. Boiler house
16. Whim rope
17. Small headframe on shaft

**Figure 31. Remains of a wheelbarrow in the adit near Old Engine Shaft. Photo: Kevin Baker.**

landed on the platform to assess the mine in 2006 before the site became part of the World Heritage Site in that year.

The two 'engine shafts' are both spectacularly sited about 20m above sea-level. They are marked on the Ordnance Survey maps as 'New Engine Shaft' and 'Old Engine Shaft' respectively to the east and west. It is not known where these names came from as neither name occurs in the few extant reports. The name 'engine shaft' implies that the shaft was used for pumping, usually by steam. As the shaft called New Engine Shaft was the first reported to have an engine installed it might be expected that this would in fact be Old Engine Shaft. However, as pointed out by Adam Sharpe (Cornwall Archaeological Unit), it is possible that pumping was being carried out by other means first at the eastern shaft (the 10-inch water pressure engine was used for pumping and could have been sited here). The convention of Old Engine Shaft in the west and New Engine Shaft in the east has been followed here. In 1845 there is reference to Rogers' Engine Shaft, which could represent the renaming of Old Engine Shaft following the erection of the new engine.

Possible clarification comes from the 1880 reopening of the mine, as New Penrose, which worked the 'old Trewavas copper lode'. There are references to the 'old engine shaft', which was being used for dumping waste rock to save the cost of hoisting it to the surface.

Old Engine Shaft is in fact a pair of adjoining shafts. The northern shaft is sub-rectangular and it is suggested by the Cornwall Archaeological Unit that this is the earlier of the two. The southern shaft may well have been sunk for hoisting, possibly because the northern shaft was too narrow to allow pumping, hoisting and, presumably, a ladder-way.

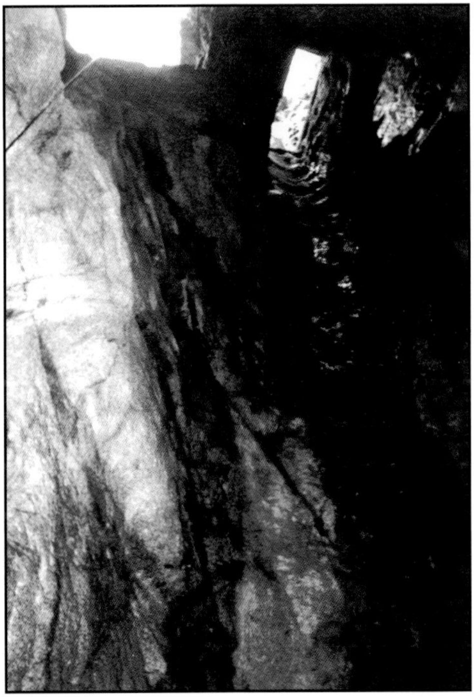

**Figure 32. Looking up at Old Engine Shaft, right, and the presumed whim shaft, left from the huge chamber below. Photo: Kevin Baker.**

New Engine Shaft is set slightly back from the cliff-edge. The ground is higher and less steep than at the Old Engine Shaft, but was cut down some 2m to accommodate the works. The rectangular shaft here has a slot for a balance bob, which was used to reduce the load on the engine by counter-balancing the weight of the pump rods in the shaft.

Ore brought to the surface at New Engine Shaft for dressing was hauled up the coastal slope by inclined tramways. Three are visible, each a straight cutting, with a platform of the type used for a horse whim above it. The easternmost cuts one on its west; it may have been replaced by the third following its reuse for flat-rods.

The tramway extending WNW of the western engine house leads to a flat area where the dressing or spalling floors were located.

> The stamps floors were of large extent; and bal-girls from Crowan, a distance of half a dozen miles, used to 'come and go' besides doing their day's work. (*Cornishman* 15.7.1880 p8)

This is the area where copper ore was 'dressed' for sale. Copper ore was never successfully smelted in Cornwall (it requires seven tons of coal to smelt one ton of ore) so dressing only involved the removal of waste material. This was carried out manually by many men, women and children breaking the ore with hammers of various weights to allow the removal of the waste rock. The idea was to produce the equivalents of the heads and tails produced by tin mines, in this case the terms 'crop' and 'halvens' were used respectively. These processes included 'spalling', 'cobbing' and 'bucking'. The work was carried out on a cobbled surface known as a spalling floor. Unfortunately for the workers many of these processes were carried out in the open, regardless of the weather, although some work was carried out in sheds. On some mines the halvens were reworked, either directly by the mine or by miners on tribute.

The ore was washed ('riddled' and 'jigged') to remove very small particles. It was then sorted into piles which were inspected by the agents of the various copper companies who would then put a price on them. It seems to have been fairly common practice for the agents to collude on the prices paid to the mine, to keep them low. It was fairly common for more than one company to buy one lot of ore as they would be looking to acquire a particular grade for their smelting works.

The equipment used for dressing copper ore (as well as for lead ores) was in the main constructed of timber. As a result of opportunistic reclaiming and recycling of these materials, it is often impossible to locate the sites of the spalling floors. Occasionally, on mines such as Botallack, the sites can be recognised by the damage done to plant growth caused by copper contamination in the soil.

Two more engine houses are known from the mine and these may possibly be seen in figures 29 and 30; however neither is shown in sufficient detail to be absolutely certain. That just to the west of the count house was probably for pumping; it has the look of an all-indoors engine house similar to that at Levant. There is no obvious protruding beam, or a bob wall. This house is not mentioned in any of the mine reports and must therefore have been built, and become disused, prior to the mine opening in 1834. The shaft it pumped from was probably the one immediately to the west. The figures also show what appears to be a whim above and slightly to the north-east of the 45-inch engine, although it is hidden behind two wheels; these could be flywheels. This is presumably the 16-inch whim engine offered for sale in October 1844, the only known reference to this engine so far located. The figures imply that it was hoisting from New Engine Shaft and probably only post-dates the 45-inch engine house by a short period.

The 1844 reference to the 'East Whim' implies that two whims were operating at the same time, however there is no evidence that both were steam-powered. When the 70-inch pumping engine was put up for sale an 18-inch whim engine was included. As with the 16-inch, there are no other references; this is either a mistake or because the 16-inch had been recylindered. The sale notice for the two engines only mentions that the whim had two boilers, one of which was likely to be a spare.

The count house, the mine's offices, lay immediately to the north of the break on the edge of the coastal slope. This two-storey building had three chimneys. Two small single-storey buildings were sited to the north-east of the count house.

As was common in so many mines, many of the built features of the mine, such as smiths' and carpenters' shops and materials stores, seem to have been cleared away in a few years following its closure. The more temporary structures, the horse whims, tramways, timberwork, head frames, shears, pitwork (pump rods, pumps &c.), engines, boilers and the re-usable roofing materials would have been amongst the first to go, leaving the buildings as empty shells open to the weather. Capstans and whim plats, yards and trackways would gradually have grown over, while the more lightly-built structures such as the boiler houses, stores, etc., and the like) would have gradually succumbed to stone robbing, with the final traces being removed

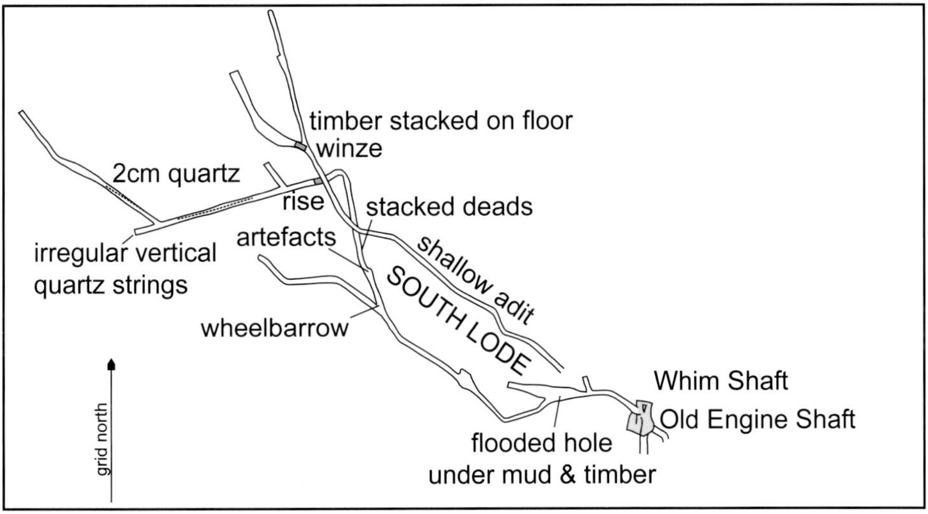


**Figure 33. Plan of adits at Old Engine Shaft surveyed by Ali Neill and Kevin Baker. Scale not known.**

by the wind and rain. The more lightly-built structures doubtless provided material for improving and repairing local homes for some years following closure.

By the late 19th century the only features left to be recorded by the Ordnance Survey comprised the two cliff-side engine houses and their stacks, the count house and the possible engine house and stack to the west of the former. By the time of the 2nd Ordnance Survey, about 20 years later, only the two engine houses on the cliffs remained to be recorded. These doubtless survived because of their locations and more robust construction.

Summary of building details

| **Feature** | **HER No.** | **NGR** | **Listing No.** |
|---|---|---|---|
| Eastern pumping engine house | 29280.20 | SW 6006 2658 | 65723 |
| Chimney of eastern pumping engine house | 29280.40 | SW 6008 2659 | 65724 |
| Western pumping engine house | 29280.10 | SW 5995 2650 | 65721 |
| Capstan platform | 29280.30 | SW 5993 2647 | 65721 |
| Chimney | 29280.50 | SW 5991 2652 | 65722 |

HER: Historic Environment Record
Listing No.: from Historic England's *Statutory List of Buildings of Special Architectural or Historic Interest*

**Bibliography**


Sharpe, A., 2008. Historic Environment Services (Projects). *Mount's Bay: Trenow, Cudden Point, Sydney Cove, Lesceave Cliff, Rinsey East Cliff and Trewavas, Cornwall.* Archaeological assessment for the National Trust. Cornwall County Council.

Sharpe, A., 2008. Historic Environment Services (Projects). *Trewavas, Cornwall.* Conservation Management Scheme. Cornwall County Council.

Sharpe, A., 2009. Historic Environment Services (Projects). *Trewavas, Breage.* Cornwall Historic Environment Consultancy and watching brief during conservation works. Cornwall County Council.

Sharpe, A., 2017. Historic Environment Projects. *Trewavas, Breage, Cornwall.* Conservation Management Plan.

Holmes, L., 2003. Discussion document on Wheal Trewavas.

# Chapter 4

## Conservation and Consolidation Work

The Wheal Trewavas site, covering 8.25 Ha, was acquired by the National Trust early in 2008. It is part of the National Trust Mounts Bay Properties, which presently comprises Trenow (Marazion), Cudden Point, Sydney Cove (Praa Sands), Lesceave Cliff, Rinsey East Cliff and Trewavas Cliff. The importance of the site is demonstrated by the fact that it is now a Scheduled Ancient Monument; it also forms an important part of the Cornish Mining World Heritage Site.

It had been long recognised that the buildings were in urgent need of conservation and consolidation works if they were not to suffer damage which could have resulted in loss of structural integrity and then collapse. A range of works was required to achieve this, and plans were put in place to carry out the necessary work. Because of the site's importance a range of reports was required to be made prior to any work being carried out on the site; these included archaeology, environment, biological, mining and contamination as well as detailed surveys of the buildings. It was also important that access to the site by local contractors did not impact on it. The protected nature of the site meant that several applications for Scheduled Monument Consent were required in order to carry out all of the work. Part of the planning process was to decide on the types of materials which would be used for the masonry work, such as the composition of the mortar.

The work carried out at Trewavas was one of the most challenging mine building conservation projects undertaken by the National Trust, requiring innovative and imaginative approaches to some elements of the works, in particular the provision of a means for delivering scaffolding and other materials to the works sites, and for providing the team of masons with a safe means of access to the base of the cliff slope. It should be noted that it was a condition of the Scheduled Monument Consent that any scaffolding could not be in contact with any of the engine house walls. Given the presence of three open mine shafts and a considerable drop into the sea, the potential dangers can easily be guessed at.

The engine houses were in as poor a condition as might be expected, bearing in mind they had been abandoned for 160 years. Both had been damaged during the removal of the engines, and both had been subjected to the vagaries of Cornish weather. A particular problem, as with many other engine houses, was the deterioration of the various wooden components, all of which were structural, which had lead to instability of the masonry. Conservation work required the replacement of the rotten and missing timbers together with repointing of the masonry. The word 'repointing' however does not really do justice to the work which was required here, not just the replacement of mortar but the sealing of walls to protect against rainwater ingress amongst other things. In addition precautions had to be taken to leave the masonry so that standing pools of water could not form at the wall heads. An important constraint, outside of the control of the contractors, was the presence of Choughs within the area; no work was to interfere with the birds during the nesting season.

The western engine house was the first structure at Trewavas to be conserved; work started in October 2008 and was completed in January 2009. This house was in a particularly poor condition: two massive lintels above the cylinder door had been lost and the third was no longer providing support; masonry above the opening was being lost; significant cracks had appeared and the tops of the rear corners of the wing walls had moved inwards towards one another. It was therefore essential that temporary stabilisation was put in place prior to any repair work being carried out.

Work on the eastern engine house was particularly complicated, starting with access. An access route was cleared of its vegetation down to the west engine house; dense blackthorn was a severe impediment. Once this was finished, a series of paired ground anchors was drilled down the incline to a point nearly 10m above ground level at the rear of the engine house. These secured an inclined scaffold; where this reached a point opposite the engine house a scaffold bridge was constructed to access the house.

The scaffolding required for this house was also complex as the presence of a double shaft in front of the bob wall meant that there could be no footings. The resultant scaffold design incorporated a number of lattice beams, while the forward section was cantilevered off the rear section and the bridge. Further beams through the ground floor openings at the front of the house, external to the bob wall, tied the scaffolding together.

A week was required to strip the scaffolding from the west engine house and re-erect on the east engine house.

The stack and stack stump were much easier to work on, and required no complicated scaffolding or ground clearance work.

**Bibliography**

Sharpe, A., 2008. Historic Environment Services (Projects). *Mount's Bay: Trenow, Cudden Point, Sydney Cove, Lesceave Cliff, Rinsey East Cliff and Trewavas, Cornwall.* Archaeological assessment for the National Trust. Cornwall County Council.

Sharpe, A., 2008. Historic Environment Services (Projects). *Trewavas, Cornwall.* Conservation Management Scheme. Cornwall County Council.

Sharpe, A., 2009. Historic Environment Services (Projects). *Trewavas, Breage.* Cornwall Historic Environment Consultancy and watching brief during conservation works. Cornwall County Council.

Sharpe, A., 2017. Historic Environment Projects. *Trewavas, Breage, Cornwall.* Conservation Management Plan.